JOURNEY of the JIHADIST

ALSO BY FAWAZ A. GERGES

The Far Enemy:
Why Jihad Went Global

America and Political Islam:
Clash of Cultures or Clash of Interests?

FAWAZ A. GERGES

JOURNEY

of the

JIHADIST

INSIDE MUSLIM MILITANCY

A HARVEST BOOK

HARCOURT, INC.

Orlando   Austin   New York   San Diego   Toronto   London

www.HarcourtBooks.com

Library of Congress Cataloging-in-Publication Data
Gerges, Fawaz A., 1958–
Journey of the jihadist: inside Muslim militancy/
Fawaz A. Gerges.—1st ed.
p. cm.
Includes bibliographical references and index.
1. Terrorism—Religious aspects—Islam. 2. Jihad.
3. Islamic fundamentalism. I. Title.
BP190.5.T47G47 2006
322.4'2088297—dc22 2005037759
ISBN 978-0-15-101213-8
ISBN 978-0-15-603170-7 (pbk.)

Text set in Book Antiqua
Designed by April Ward

Printed in the United States of America

First Harvest edition 2007
A C E G I K J H F D B

For Nora, Bassam, Annie-Marie,
Hannah, and Laith

Contents

PROLOGUE

SEVERAL MONTHS before September 11, 2001, an Al Qaeda recruitment videotape was smuggled out of Afghanistan and circulated clandestinely in the Arab world (it had not been intended for public view). In it, Osama bin Laden and his senior aides lecture young Muslim men about their duty to Islam by referring to the exodus or the journey from Mecca to Medina that the Prophet Mohammed and his *sahaba*, "companions," made in 622 C.E. "Having diagnosed the disease," says bin Laden, after a diatribe about the encroaching evils of the West, "here is the medicine. And the cure is Allah's Book: hijra and jihad."

Hijra refers to a journey from a land where Muslims cannot practice their *deen*, "religion," to a place where they can. In the military sense that bin Laden intended here *jihad* means fighting in the cause of Allah. Mohammed had spent thirteen years trying to institute Islam in Mecca, where he and his sahaba were persecuted; an attempt was even made on the Prophet's life. He had a *wahi*, "revelation," that immigrating to Medina would not only escape persecution but also transform the world. Under cover of darkness, and pursued by Meccans, pagans (Jahili) who did not believe in Islam, he and his followers made the journey to Medina.

The source of hijra relates to this journey, a concept of profound spiritual significance to Muslims. Bin Laden would have his listeners believe that the jihad meant the Prophet's eventual return to Mecca, cleansing it of unbelievers after defeating the Meccans at the battle of Badr in 624. Bin Laden would also have his listeners believe that his forced exodus from Saudi Arabia (first to Sudan and then to Afghanistan) in order to wage jihad against America and its allies parallels the Prophet's hijra to Medina. Like Mohammed, bin Laden and his cohorts would suffer and make sacrifices. But, since they were following God's will and the injunctions of Mohammed and the "righteous predecessors" (*salaf*), they would eventually prevail. Afghanistan was the Medina of the twenty-first century.

In the Qur'an, however, formal sanction to fight—to wage jihad—first finds expression in the so-called Medinan Period: "Permission to fight is given to those [believers] who are fighting them [unbelievers] [and] because they have been wronged, Allah is able to give them victory. Those who have been expelled from their homes unjustly only because they said: 'Our Lord is Allah'" (22:39–40). In Medina, Mohammed had further revelations about *gital* (fighting). It was obligatory ("Fighting is ordained for you though you dislike

2

it, and it may be that you dislike a thing which is good for you and that you like a thing which is bad for you. Allah knows but you do not know" (2:216); it was a form of *tawba*, "repentance"; and it was a covenant with God ("Allah hath purchased of the believers their persons and their goods; for theirs [in return] is the garden: they fight in His cause, and slay and are slain: a promise binding on Him in truth, through the Law, the Gospel, and the Qur'an: and who is more faithful to his covenant than Allah."). Sanction became duty. The Prophet was ordered by the Qur'an to encourage the believers on the path of jihad: "Then [O Mohammed] fight in Allah's cause. Thou art held responsible only for thyself—and rouse the believers. It may be that Allah will restrain the fury of the Unbelievers; for Allah is the strongest in might and in punishment" (4:84).

The journey of the jihadist was transformed into an open-ended quest, a path that all Muslims needed to embark upon to emulate the Prophet of Islam. Bin Laden believed that his reference to the Prophet's journey or exodus from Mecca to Medina would resonate in the hearts and minds of those upon whom he was now calling.

But how deeply did it resonate? For more than three decades, an internal struggle has been waged for the soul of Islam, a struggle that has shaken the very foundation of Muslim societies and politics. Its reverberations have been felt far beyond the region's borders—in New York, Washington, Madrid, London, and Paris. The primary goal of modern jihadism is and always has been the destruction of the atheist political and social order at home and its replacement with authentic Islamic states. But since the late 1990s these jihadists have fought bitterly among themselves. It was then that bin Laden and his second in command, Ayman al-Zawahiri, launched a campaign to hijack the movement and change its direction—away from attacking *al-Adou al-Qareeb*, the "near enemy" (Muslim "apostates" and "renegades") and toward attacking *al-Adou al-Baeed*, or the "far enemy" (Israel and the Western powers, particularly the United States). By taking on the United States, which in the eyes of so many Muslims is most responsible for maintaining the grim status quo in the Arab world, Al Qaeda jihadists wanted to signal that the civil war for the soul of Islam would move to a different level, an international stage. From their perspective this could further two goals: rid Muslim countries of corrupting American and Western cultural and political influences, as well as military presence, particularly from Saudi Arabia, the birthplace of Prophet Mohammed; and destabilize Muslim governments and their ruling elite by inciting the rising generation.

September 11 was bin Laden's attempt to turn the wheels of political fortune in his favor—by proving to the Muslim world that he and his brethren now represented the vanguard of the *ummah*, "Muslim community." He believed that the very outrageous boldness of the attacks would attract new recruits. But the success of the plan depended entirely upon America's response. Were the United States to expand its war against Al Qaeda into countries that had nothing to do with the attacks, bin Laden knew that fellow Muslims would react with anger—with jihad.

The vast majority of Arab activists did not join Al Qaeda.

In fact, September 11 showed how deep the fissures within the global Muslim community ran. The Western media still perpetuate the myth that the attacks were widely embraced by the ummah, that the nineteen suicide bombers reflected the amorality of Muslim political culture as a whole. Suicide bombings and beheadings sustain this myth and rekindle deep Islamophobic tendencies. Jihadism has fed the assumption that Islam and Muslims embrace and indeed celebrate a culture of death.

Born and raised a Christian in Lebanon, I grew up in a culturally diverse Arab society, among Muslims and Christians both. We celebrated religious holidays together. We shared hopes, dreams, and fears. I still travel frequently to Arab countries, where I've spent years living and working with Muslims, and have come to understand how misrepresented and misunderstood they are in the West. It is a terrible time to be a young Muslim. Most feel profoundly depressed, incapable of realizing even their simplest aspirations. They are politically oppressed and socially repressed in their countries of birth, unable to find jobs so they can afford to rent apartments or even get married. Abroad they face racial profiling, perceived as carriers of a plague of nihilism that must be kept in isolation. Most are denied visas to Western countries, particularly the United States, to study and work. Welcome nowhere, they have become the pariahs of the twenty-first century.

These young Muslims cannot understand why they are being punished for the crimes they have not committed and do not condone. "Does not collective punishment violate the very principles and values of liberal democracies?" they often inquire. Muslims under the age of thirty make up about 60 percent of the world's Islamic population; they represent a huge, disfranchised constituency. Yet only a tiny fraction turns to violence and terrorism.

The Arab world did not invent terrorism and has no monopoly on this deadly trade. In decades past, homegrown militant factions of various ideological orientations have terrorized European societies and caused hundreds of civilian deaths and injuries. Yet, even at its height, no pundit linked European terrorism with a crisis in Western culture and religion, or muttered darkly about the decay or corruption of the Christian West.

Some commentators have even argued that Al Qaeda's terror tactics are grounded not in the Islamic tradition of jihad but in far more recent European, radical, ultraleftist, or Third Worldist movements. From that point of view, Al Qaeda represents a rupture within a religious culture that does not condone nihilistic offshoots. Al Qaeda has only been couching its political struggle in religious terms because it wants to convince the ummah that it is fighting on its behalf. That is why bin Laden reprimands fellow Muslims for neglecting jihad: "No other priority, except faith, could be considered before [it]. In fact, it represents the pinnacle of religion. How can religion survive without it? Our nation's life, glory, and survival are at stake. Our enemy tells lies; our religion tells the truth when it tells Muslims: 'If you do not punish the Crusaders for their sins in Jerusalem and Iraq, they will defeat you because of your failure. They will also rob you of the land of the Two

Holy Sanctuaries [Saudi Arabia]. Today [they robbed you] of Baghdad, and tomorrow they will rob you of Riyadh.'"

I do not presume to claim that the East is morally superior to the West, or that religious and ethnic minorities have lived in perpetual peace in the Muslim world. Such claims would not be true. The demise of the Ottoman Empire unleashed social and political upheaval and widespread persecution of minorities. But Britain and France's division of the region into separate spheres of influence and the subsequent emergence of nationalist states planted the seeds for today's struggles, including the Palestinian-Israeli conflict. The roots of anti-Semitism are European. Anti-Jewish sentiments in Muslim countries are a result of the creation of the state of Israel and the displacement of millions of Palestinians. European anti-Semitic tracts (*The Protocols of the Elders of Zion* and works by Holocaust-denying writers) still provide the main diet for racist Muslims. I do not mean to rationalize or minimize anti-Jewish feelings in Muslim societies but to dispel the popular myth that situates them in religion rather than politics. Racial anti-Semitism has not migrated from Europe into Muslim lands. After all, both Arabs and Jews are Semite and share a common history of persecution. Being labeled anti-Semite is ludicrous to most Arabs. "How could we get out of our own skin?" a leading Islamist, Laith Shubialat of Jordan, once asked.

There is nothing unique about the social and political turmoil roiling the Arab world, though conventional wisdom holds that Muslims are inherently volatile. Our media send that message every day. There is no denying that political violence seems to thrive in the Middle East (which, incidentally, is not synonymous with Islam or with Muslims; more Muslims live in one non–Middle Eastern country, Indonesia, than in the entire Arab arena, in which there are also sizable Jewish and Christian communities). Suicide bombs reinforce perceptions about the savagery of the struggles there. The world economy runs on cheap Arab oil, intensifying Western fears about the repercussions of political instability in the region. Yet in comparative historical terms, the Muslim world is not as violent as headline news would have us believe. The Arab political order is very young. We cannot blame colonialism for the extremism and violence that persist in the Muslim arena, but the die was cast a century ago, when local aspirations of freedom and self-determination were sacrificed at the colonial altar.

Fragile or not, the Western-style political system has taken root in Muslim soil. National boundaries are unlikely to be redrawn. Local rulers have vested interests in maintaining existing political arrangements. Likewise, the great powers have too much to lose. When Saddam Hussein invaded neighboring Kuwait in 1990 and annexed it with Iraq, the United States deployed 500,000 troops to the Persian Gulf. The world's great powers end up benefiting considerably from the region's political instability. The despotic rulers who have been put in place depend on the West for survival and, as a result, tend to be pliant and accommodating. But September 11 showed that this arrangement comes at great cost.

Even without colonial intervention, nation building tends to be a prolonged, complex, and costly process. My family and I were witnesses to, participants in, and victims of the violent sociopolitical evolution that began with the Lebanese civil war in the 1970s and continues to this day. But painful as they are, personal and psychological factors must not overshadow historical truths. Their cultures may be ancient, but Lebanon and other Muslim countries fit the historical pattern of sociopolitical development. Many nations, including the United States, experienced bloody sectarian strife before evolving into stable democracies. What is different is the religious rhetorical framework within which this evolution is taking place.

Religion has emerged as a powerful tool of dissent because authoritarian Muslim rulers succeeded in silencing secular, progressive, and nonreligious opposition. One Egyptian militant boasted that "today secularists, Marxists, and nationalists represent a big zero in the Arab political equation." Even Muslim dictators have been unable to control the mosque, which has taken the place of mainstream civil society and provides a safe haven for the voices of the dissatisfied.

But mainstream Islamists, not jihadists or liberal democrats, were the direct beneficiaries of these limited democratic openings. In the last several years, Islamic activists scored impressive electoral victories in Egypt, Iraq, Iran, Lebanon, Saudi Arabia, Morocco, Pakistan, Kuwait, and Turkey. In January 2006, world leaders were caught off guard by the spectacular victory of Hamas (whose official name is the Islamic Resistance Movement) in the Palestinian parliamentary elections. Winning 74 out of 132 seats, Hamas swept the governing Fatah party (which won 45) from power.

We should not be surprised that Muslim voters are empowering mainstream Islamists. Secular rulers have failed both to deliver jobs, social services, and education, and to defend the homeland against external threats. More and more Muslims view Islamists as the most effective alternative to the discredited ruling establishment. Muslims are also impressed by Islamists' incorruptibility, as well as their ability to expose the incompetence and failings of secular regimes. They put their money where their mouths are. Hamas, for instance, helps the widows of suicide bombers and provides daycare, healthcare, and early schooling for the children of the very poor.

Nonetheless, Islamists will meet a fate similar to that of their secular precursors if they do not deliver the goods or if they entrap their constituents in costly military adventures against real or imagined enemies. The Muslim electorate is not itself Islamist in a radical or fundamentalist way; rather, it is disaffected and fed up with oppression, corruption, and incompetence, and it is willing to "throw the bums out," a perfectly natural occurrence in a democratic society.

Cradles of democracy or incubators of militancy? Mosques seem to serve as both. Time and again jihadists told me they met at local underground mosques to recruit foot soldiers and discuss tactics and strategy. Almost every one found his "calling" at a specific mosque, where he made connections that proved enduring and effective. We know now that some of the September 11 suicide bombers were initiated into jihadism in militant mosques in various European cities.

Many Muslims have expressed rage at the accusation

that their mosques promoted violence. And it is true that the overwhelming majority of mosques are devoted to worship and contemplation. Nonetheless, there are plenty of clerics who have embraced political-religious dissidents and allowed their mosques to be taken over by extremists.

All of this should not blind us to the broader sociopolitical goals and ambitions that lie behind the veil of religious rhetoric. It is the weapon of choice of disaffected individuals and groups, useful against both the "near enemy" and the "far enemy." Islam confers legitimacy on those who challenge secular nationalism, which has already been discredited by decades of political oppression, economic meltdown, and military defeat. The jihadists I spoke with all referred to Israel's 1967 defeat of the Arab states as a watershed in their radicalization and revolt against "apostate" rulers. Armed with a new militant, or Islamist, perspective, they launched a frontal assault against the failed ideologies of nationalism, socialism, communism, and "decadent" Western liberalism in a daring bid to overthrow Muslim rulers and establish Islamic states. Religion became their tool for political mobilization.

Thus the key to understanding the jihadist and his journey lies in politics, not in religion. Blaming terrorism on passages from the Qur'an would be like blaming the Crusades on passages from the New Testament. As is true for the Bible, Islamic doctrine can be interpreted in any number of ways, either to promote peace and tolerance or war and intolerance. The real culprits are the ideologues who would twist religion—any religion—to serve their political agenda. The challenge is to decipher what Muslims say and do rather than get bogged down talking about Islamic or Qur'anic doctrine.

Studying Islam itself will not help us understand the goals and strategies of the jihadist. Muslims and "Islamists" do not agree on what Islam is, let alone its role and function in governing their lives. Muslims are believers who may or may not be interested in politics, while Islamists are political activists whose fundamental goal is to seize power and Islamize state and society, preferably from the top down. Just as there are many kinds of jihadists, however, there are many kinds of Islamists—from moderate to mainstream to militant. There are different forms of Islamism, just as there are many forms of Islam—Sunni, Shiite, Salafi, Sufi, and Wahhabi being the major ones—all of which reflect different tribal (and now national) distinctions and philosophical traditions (such as which religious texts should receive greater or lesser emphasis). Palestinian Islamists are generally Sunni; Egyptian, Sunni; Saudi, a blend of Salafi and Wahhabi; Iranian, Shiite. These overlap, interact, and intersect in ways that affect alliances and perspectives. Some are more intolerant than others. They all share the goal of establishing Islamic governments based on *shariah*, Qur'anic law. Beyond that, few see eye to eye about what constitutes an "Islamic" foreign policy or an "Islamic" economic program. Their organizations possess no concrete views of politics or economics nor do they offer new models of society.

Mainstream Islamists represent an overwhelming majority of religiously oriented groups (in the upper ninetieth percentile); they accept the rules of the political game, embrace democratic principles, and oppose violence. Militant Islamists share with jihadists a willingness to use all means at their disposal, including terrorism, to overthrow the existing secular order and replace it with a theocratic one. Jihadists may have a less sophisticated perspective of the religious nature of their struggle than mainstream Islamists; they may be less concerned with nourishing their followers on a rich diet of moral rhetoric. But they share the conviction that force is justified. Militant Islamists and jihadists read from the same texts—such as the ones bin Laden was referring to in that recruitment video—and draw the same political conclusions.

⊂══◆══⊃

Beginning in 1999, I conducted scores of interviews with both militant Islamists and jihadists (henceforth, when I refer to "Islamists" it will be to the militant rather than the moderate or mainstream branches, which are not the focus of this book). While these people normally shy away from contact with the Western media, I was able to probe their views on such subjects as the role of government, the use of force at home and abroad, and economic and foreign policies. Their answers reflected a broad political spectrum, from the extreme left to the extreme right. What most surprised me was that their views tended to resemble those of their secular opponents. I have come to believe that American observers assign far too much significance to notions of "Islam" and "Islamic" when describing what motivates Islamists and jihadists. I do not mean to downplay the role of "Islamic" factors in Muslim societies, particularly in how they help shape moral values, self-identification, and fears about Westernization and Americanization. But again, both Islamists and jihadists use religion as a means to a political end, not as an end in itself.

I noted two general trends in my interviews, both deeply ironic yet instructive. The first is that while their goal is to dismantle the secular authoritarian order that succeeded British and French colonialism after World War II, Islamists and jihadists ultimately want to establish governments that would be similarly repressive. Countries that have Islamic governments, such as the Islamic Republic of Iran, the Sudan, and, formerly, Afghanistan under the Taliban, provide cases in point. Although fully clothed in Islamic dress, they have much in common with their secular authoritarian counter- parts elsewhere. There is nothing uniquely "Islamic" about their internal governing style except the rhetoric and the symbolism. They have not offered up an original model of Islamic governance. Many Iranians and Sudanese seem fed up with the economic damage and sociopolitical upheaval inflicted by the mullahs and Islamists on their societies. The signs of decay and entropy are everywhere. Not only did the Taliban fail to establish a functioning government, they obliterated the rudimentary institutions they had inherited. Their example casts a very dark shadow over the concept of an Islamic state. In foreign policy, the three Islamic republics invested their meager resources in exporting revolution to neighboring states and beyond. This ambition has proved to be costly and even suicidal for the Taliban, which hosted Al Qaeda and allowed it to wage jihad worldwide. Since September 11 the two remaining Islamic regimes in Iran and Sudan have retrenched domestically and are mainly preoccupied with political survival, though the new ultraconservative Iranian president, Mahmoud Ahmadinejad, has worked aggressively to roll the clock back to the early days of the Islamic revolution in the late 1970s.

The second irony is that Islamists and jihadists are playing an active if indirect role in expanding political debate in the Muslim world. They have forced existing secular dictatorships—such as in Egypt, Tunisia, Morocco, Algeria, Turkey, and even Saudi Arabia—to respond to their challenge to open up the closed political system and reform government institutions. Without such pressure, these authoritarian Arab rulers would have no incentive to respond to demands for inclusion and transparency. Opponents of Western-style democracy, jihadists are unwitting harbingers of democratic transformation. Some have even come to endorse a political process, creating further reformation within the movement. Al-Jama'a al-Islamiya (Egyptian Islamic Group)—the largest jihadist organization in the Arab world—and the Islamic Salvation Front (FIS) in Algeria now emphatically support democratization. The journey that led them there was far from democratic, however; it was extremely bloody. In the 1990s, aware that they would never be able to dislodge autocratic rulers by force, they began to alter their strategy for gaining power. They even formed alliances with their former sworn political opponents, including secularists and Marxists, in calling upon governments to respect human rights and the rule of law.

Islamists and jihadists are not born-again democrats and will never be. They are deeply patriarchal, seeing themselves as the guardians of faith, tradition, and authenticity. Their rhetoric remains soggy with anti-Western diatribe. Even mainstream Islamists oppose women's rights and empowerment, or freedom of expression for critics and artists. In Kuwait and Saudi Arabia, Islamists have vehemently opposed efforts to give women the right to vote or to drive cars. In Egypt, Morocco, Jordan, Tunisia, Algeria, Pakistan, and other Muslim countries, they denounce any legislation that would enable women to divorce abusive husbands, travel without male permission, or achieve full representation in parliaments and state bureaucracies.

Nonetheless, many Islamists and a large segment of jihadists are gradually becoming initiated into the culture of political realism and the art of the possible. They are learning to make compromises with secular groups and to rethink some of their absolutist positions. Events have forced them to come to grips with the complexity and diversity of Muslim societies. More and more they recognize the primacy of politics over religion and the difficulty, even futility, of establishing Islamic states, particularly by autocratic fiat. Conservative "neo-fundamentalism" (which aims primarily at Islamizing society from the bottom up through what is called *da'wa*, "the call") has generally replaced revolutionary jihadism, whose goal is to Islamize society by simply seizing state power.

But not all Islamists and jihadists have embraced these new political realities, and therein lies the internal struggle. There is a split between the ultramilitant wing, including Al Qaeda, and a nonviolent faction that commands greater numbers and political weight. This civil war has been over-shadowed by the war in Iraq, which was a godsend to Al Qaeda because it diverted attention from its zero-sum game and lent it an air of credibility. American officials now acknowledge that the war in Iraq has proved a powerful recruiting tool for Al Qaeda and given it time to regroup. However, the Iraq war has merely slowed an inevitable shift in the balance of power toward those who have abandoned Islamizing society from the top down. Most activists I spoke with recognize that the individual believer should be the focus in their efforts to create a moral society. Evolution, not revolution, is the dominant trend.

Jihadists still have a long way to go before they gain the trust of their fellow Muslims, let alone the international community, but some have taken an important first step. The terrorism perpetrated by certain cells and factions will continue over the next decade, but their movement no longer has a large base of support or a safe haven in which to plot new operations. Jihadists of all stripes know they are at a crossroads. At home and abroad they are blamed for unleashing the wrath of the United States against the ummah. Only a miracle will resuscitate jihadism. The question, of course, is whether the continued occupation of Iraq will be that miracle.

To understand this civil war for the soul of Islam as well as its tragic repercussions—including September 11, the Madrid and London bombings, the war in Afghanistan, and the insurgency in Iraq—we need to hear from those engaged in waging it. The "Arab street" so often evoked by even the best Western journalists is in great part a myth designed for Western consumption. Beneath it lies the simplistic notion that all Muslims and all Arabs speak with one voice—a voice baying for bloodshed in the name of religion. Living in the United States but having been raised in Lebanon, I am keenly aware of the distortion. Growing up, I heard a wide spectrum of Muslim voices. We need—more than ever—to hear what is being said in alleyways, cafés, apartment courtyards, barbershops, classrooms, and underground bunkers throughout the Middle East. I went back and began having conversations with activists and opinion makers, radical emirs and foot soldiers. My cultural background, educational training, and ability to speak Arabic allowed me to enter into their universe. So I asked questions and then I listened, learning to discern the diversity that lay beneath the sometimes monotonous and numbing rhetoric. As the voices in this book will attest, the jihadists' journey did not begin on any single day, nor did it end there.

I

Portrait of a Jihadist:
The First Generation

ONE NIGHT IN 1999 during Ramadan, Islam's holiest month, a time of fasting and abstinence, I found myself in Syedeh Zeinab, a historic, run-down, and always crowded neighborhood of Cairo. I was headed for a late-night meeting with a man named Kamal el-Said Habib. I knew him only by reputation. Many of those concerned about events in the Middle East did. He was the former leader of a wing within al-Jihad, "armed struggle," a paramilitary organization that had played a pivotal role in the assassination of Egyptian President Anwar Sadat in 1981. Al-Jihad's goals were chillingly straightforward: to decapitate Egypt's secular state, one of the oldest in the world, and replace it with an Islamic polity based exclusively on shariah, or Islamic religious law. Kamal was a key figure in the first generation of Muslim militants, who in the 1970s had planted the seeds of jihad throughout Muslim lands. If I wanted to locate the starting point of the jihadist movement, I needed to find Kamal el-Said Habib.

My own journey into jihadism thus began on that trip, two years before September 11 and everything that has followed it. "Jihadist" and "jihadism" had not quite yet entered the vocabulary of ordinary Americans, but I was determined to find out more about what was happening outside of most Western perspectives. However, once I had finally managed to arrange a meeting with Kamal, his friends and associates cautioned me. The man I would encounter, they said, was no longer the firebrand who once had dedicated his life to turning the world upside down. He and his generation had learned the hard way what happened when you tried to Islamize society through force. The bloodshed that had resulted was horrific; there were executions and lengthy prison sentences (Kamal had spent ten years in an Egyptian prison); families were destroyed. He and other former "warriors of God," had now, I was told, reassessed their old ways. They were charting a new course, one committed not to violent revolution but to political persuasion and *da'wa,* or religious calling. They were older now. They were wiser.

What I hoped to learn from Kamal was how deeply this reinvention had taken hold. I found myself fascinated by his story, for it seemed simultaneously a reflection of and contradiction of what was happening in the Arab world.

Kamal was born to a lower-middle-class family in Deemshalt, a large village with a population of thirty thousand located in Markaz Dikranis, in the Daqahlyya province in the Nile delta, sixty or so miles north of Cairo. His father traded in agricultural products, catering to the needs of the villagers who depended on seasonal agriculture for their livings. Kamal graduated from Cairo University in 1979 at the top of his class, with a degree in political science. Of his nine brothers and sisters, Kamal's future seemed the most assured. Charismatic and ambitious, he could have trained to become an academic, a writer, or a lawyer, rising above the comfortable though fairly modest standard of living of his parents and grandparents. But neither money nor the allure of ascending the ladder in *jahili,* "un-Islamic," institutions appealed to him. When he was called up for military service, he rejected the rank of a junior officer, preferring to remain a simple soldier. His goal was to replace *jahiliya* with *hakimiya,* "God's sovereignty," to bring an end to Egypt's moral and social decline.

By the time he graduated, Kamal and other Islamic activists were prepared to rise up in arms against *kufr,* "unbelief," and from the late 1970s until the end of the 1990s they waged an all-out struggle. They saw themselves as a vanguard that would restore Islam as a complete way of life. Thanks to Kamal's al-Jihad, renamed Tanzim al-Jihad, and al-Jama'a al-Islamiya, the largest jihadist organization in the Muslim world, a devastating war took place inside Egypt between 1992 and 1997. Though little recognized by the outside world, it caused billions of dollars in damage to the economy and resulted in thousands of deaths. By the end of the 1990s the war was over; the Egyptian government had won. Security services in Egypt and Algeria had killed as many as ten thousand militants. Tens of thousands more were imprisoned, their families and sympathizers fired from their jobs and denied even the most minimal social support. Their mosques, bookstores, publishing offices, and papers had been closed. Most conceded defeat and declared a unilateral "ceasefire," a code word for surrender. Throughout the 1990s Kamal was a voice of moderation, calling on his brethren and former colleagues to lay down their arms and negotiate with the local authorities.

A determined minority chose to keep fighting. But rather than continue their resistance against the entrenched Arab governments, Ayman Zawahiri, Osama bin Laden's deputy, took the fight outside Egypt. Bin Laden and Zawahiri declared war against the world's last remaining superpower, the United States, and its allies, focusing upon the "the head of the snake," the "Great Satan" itself, with the hope of resurrecting militant jihadism among the rank and file. Bin Laden's Al Qaeda was therefore at odds with Kamal and with the great majority of jihadists, who had by then rejected violence as a means of gaining political power. The jihadist civil war had started, one that would spread throughout the Muslim world and determine the future of the entire movement.

It is not by happenstance that it all started in Egypt. With seventy million people the most populous Arab state and the Muslim world's cultural and intellectual epicenter, Egypt was the birthplace of the modern jihadist movement, and remains to this day the best place to understand its complexities and locate its fault lines. The movement's founding fathers were almost entirely Egyptian, as were the authors of many of its defining documents. At its height, al-Jama'a al-Islamiya's membership was estimated to be in the tens of thousands. Along with Zawahiri, Egyptians—Abu Ubaidah al-Banshiri and Mohammed Atef (also known as Abu Hafs al-Masri)—had founded Al Qaeda with bin Laden. Al Qaeda's *shura*, "ruling council," which constituted bin Laden's inner circle, was dominated by Egyptians, who supplied organizational skills and military expertise.

Egypt is also home to the first mainstream Islamist organization, Ikhwan al-Muslimun, the Muslim Brotherhood, founded in 1928 by Hassan al-Banna, a school teacher turned ideologue. Initially, al-Banna, whose outlook was permeated with the teachings of Sufism—a philosophical, mystical, and inner-directed school of Muslim thought—intended the Brotherhood to be a youth organization aimed at moral and social reform through education and propaganda. However, it soon evolved into the largest political opposition group in Egypt. Al-Banna, a student of Rashid Rida, a well-known early-twentieth century Egyptian Islamic reformer, called for the revival and reformation of Islam and the restoration of broken links between tradition and modernity. In 1940 he defined the Brotherhood's mission in the broadest possible terms as embracing "a Salafiya message, a Sunni way, and a Sufi truth," and saw it as "a political organization, an athletic group, a scientific and cultural union, an economic enterprise, and a social idea." At the heart of al-Banna's vision was the belief in moral regeneration as a stepping stone to Islamizing state and society—in other words, it would occur from the bottom up rather than the top down, which Islamists and jihadists favor. His approach to Islamic revival was grassroots in nature and philosophy.

By the end of World War II, however, al-Banna's original vision had been dramatically transformed; the Brotherhood carried out attacks against the British military and aggressively campaigned against social and political injustice at home. In 1948 it joined the Palestinians in the war against Israel and distinguished itself on the battlefield. A number of Egyptian officers, including two future presidents, Gamal Abdul Nasser and Anwar Sadat, developed their ideology during the Palestine war and established informal links with the Brotherhood, whose anti-imperial stance won it public sympathy and support. By the end of the 1940s, the Brotherhood had developed a formidable political machine, and thought to have almost a million members.

The Brotherhood's ambitions swelled with its ranks. Some elements flirted with violence, establishing al-Jihaz al-Sirri, "secret apparatus," an underground paramilitary unit within the political wing. The Egyptian government decided things had gone too far. In December 1948 it banned the Brotherhood, accusing it of being involved in the death of Prime Minister Mahmud Fahmi Nokrashi. On February 12, 1949, al-Banna himself was shot while getting into a taxi and died a few minutes later in a nearby hospital. Later, evidence surfaced that suggested that his assassination had been planned, or at the very least condoned, by the government. The murder of the moderate al-Banna radicalized the

Islamist movement and set the stage for more violence and bloodshed. Many of the subsequent activists, including the spiritual father of modern jihadism, Sayyid Qutb, emerged from under the cloak of the new radicalized Brotherhood. Thus, directly and indirectly, the Brotherhood also shaped Kamal and his generation.

Beginning in the early 1970s, the Brotherhood—with branches in the Middle East and in Central, South, and Southeast Asia—moved more and more into the political mainstream, renouncing force or the threat of force to attain its goals. It has stayed more or less on that path, seeking to participate in the process as a legitimate opposition party, though often excluded from political participation by the ruling autocrats. "We live in a very repressive atmosphere," said Mohammed Habib, the Brotherhood's second in command. "We are keen to form our own political party. The regime won't allow us," he told the *New York Times* in August 2005. A few months later, during the November parliamentary elections, Habib sounded a little more hopeful. "These elections are different from the previous ones, where there was a large degree of repression, confinement, and pursuit, and up to 6,000 of us were imprisoned." For the first time, the Egyptian government allowed the Brotherhood, which remains banned, to take a more prominent role in promoting the candidates it endorses. In the November 2005 parliamentary elections, the Brotherhood even put women on their ticket, a dramatic move for a patriarchal Islamist organization, even a mainstream one.

Despite restrictions on voting and the arrest of thousands of its supporters, the Brotherhood won 88 out of the 444 parliamentary seats, making it the largest opposition force in Egypt (the secular opposition parties won only a handful of seats). Its electoral performance was all the more impressive because the Brotherhood had contested only 120 seats. The results caught the Mubarak government and its National Democratic Pary (NDP) off-guard. Were free and open elections held in Egypt today, the Brotherhood would win a comfortable majority, given its popularity among the middle class.

⊕═══╬═══⊕

When I arrived in Cairo in 1999, the heated tension between Kamal and his generation on the one hand, and Zawahiri and the so-called Afghan Arabs on the other, had reached a critical juncture. Kamal's ideologically softer tones were being drowned out by threats from Al Qaeda (which also called itself "World Islamic Front for Jihad Against the Jews and Crusaders").

Kamal had not been especially forthcoming during the many calls I had to make before persuading him to let me interview him. "Why are you interested in interviewing me?" he had immediately asked. His suspiciousness was not surprising. All jihadists and Islamists believe that non-Muslims interested in them are either hostile to their project or are outright CIA spies. These are not baseless assumptions. For decades the CIA recruited journalists and scholars across the Arab world, a practice suspended in the 1970s but then resumed after September 11. Conspiracy theories thrive in Arab politics; American and Zionist agents are believed to be everywhere. Every American scholar, every journalist, is regarded as a potential spy, guilty until proven innocent.

Moreover, Kamal had spent ten years in Egyptian prisons, not known for their humanitarian treatment of inmates. While there, he had composed an essay in which he argued that Western interest in the Islamist movement had two bases: "First—antipathy, jealousy, hatred. Second—fear, apprehension, and terror.... The nature of the struggle between Islam and Christianity and Judaism is civilizational," he had written, "and thus it is existential. This struggle will not end unless one opponent defeats or annihilates the other, or forces him into his circle of influence by forsaking his faith." In the younger Kamal's eyes, we were all foot soldiers in the war between Islam and the Christian West; there could be no neutral observers or internal dissidents. Softened or not, those views still formed the philosophical platform of all modern jihadists.

But that had been in 1986, at the height of another revolutionary era. After being released from prison in 1993, Kamal kept his distance from all organizations, even the one he had helped to found, now preaching as a committed "son of the movement," as he described himself. Consequently he had gained credibility with both sides in the civil war, the confrontationist as well as the accommodationist camps; his views were respected even by those who disagreed with him. "I do not represent Islamists," he would later tell me. "I hold an independent position. I influence *al-shabab* around me without belonging to a particular organization." *Al-shabab* means "youth" and is a word used in everything from teen magazines to soccer leagues. To Kamal, however, it represented the vanguard of young believers, the youthful purpose of a rising generation.

Could a jihadist temper his views and still be jihadist? It was one of the questions I wanted to ask Kamal, and why I kept calling him to try and set up an interview. I wanted him to teach me, I told him. I wanted to learn about the transformation within the movement over the last two decades.

Finally, almost resignedly, one day Kamal suggested that I come to his office in Syedeh Zeinab that same evening. He was working as a writer and researcher at *Al-Shaab,* a daily paper that was the official mouthpiece of the Islamist Labor Party. After I hung up I remembered that Egyptian colleagues had strongly advised me to meet with Kamal during the daytime, and only in public spaces.

As my taxi pushed through waves of pedestrians on its way to Kamal's office for our first face-to-face meeting, I could see that Ramadan had put Cairo in a festive mood. Everywhere were lights and vendors and shoppers engaged in impassioned haggling. I rolled my window down. I felt the urge to plunge into the crowds, into the season's communal rituals. This was the culture I had grown up with in Lebanon; these were the sights and sounds that made me feel at home in the world. The driver deposited me in front of a dreary concrete structure. Climbing up the stairs into the darkened building, I fought a childish temptation to turn around and leave.

Kamal's office was on the second floor. I was told to wait in the tiny reception area until he had finished the last of his five daily prayers. The splotchy white walls were bare except for a few posters bearing beautifully lettered Qur'anic verses. I sat in one of the two plastic chairs. An electric bulb hung from the ceiling. A young man with a scraggly beard sat at a battered metal desk, sorting through a stack of typewritten pages that would, I assumed, go into the next day's edition of *Al-Shaab*. Day in, day out, the paper's writers and editors, all radical Islamists or Islamist sympathizers, hammered away at what they regarded as the decadent Egyptian regime. Kamal wrote about everything from Muslim minorities in the Americas to spiritual issues of the day to the suffering of Palestinians living under occupation. *Al-Shaab*'s readers had grown used to the paper's persistent financial troubles and publication delays, given that its editors were continually harassed and occasionally imprisoned.

Fifteen minutes later, a well-built but slightly stocky man with a thick graying beard emerged. Kamal looked much older than his forty-two years. The ten he had spent in prison were written all over his pale, wrinkled face; his eyes, though still bright, were tucked inside sagging lids. His hands were scarred—from cigarette burns inflicted in prison, he told me later. He wore a white shirt open at the collar and light blue trousers. He came straight toward me with a steady stride and shook my hand.

"Nervous about meeting a former terrorist?" he asked as he led me into his office.

That first time, we talked for two hours. At first Kamal
shied away from controversial or personal questions. He
clearly had reservations about opening up to a foreign academic and writer. But the fact that we could converse in
Arabic made an enormous difference. The more I impressed
upon him my determination to present the human dimensions and motivations of the jihadist movement, the less
suspect I became in his eyes. I reassured him again and
again that I had no hidden agenda, nor any relationship
with any government anywhere. Slowly, his suspicions began to evaporate.

Over the course of the next few years, Kamal came to
trust me. He even took it upon himself to introduce me to
other former activists of al-Jihad and al-Jama'a al-Islamiya.
After a point, Kamal no longer needed prompting from me
to open up about his life; he was even anxious to tell his
story, a story that in his opinion was being distorted by the
propaganda wings of Egyptian President Hosni Mubarak's
government and its vast network of informers and agents.

He began with his student days at Cairo University,
which he entered in 1975. Like Kamal, many of the under-
graduates were the first in their families to pursue any kind
of higher education. The university was then home to
150,000 students, who stuffed themselves into lecture halls
still named for the heroes of British civilization (today, the
buildings are called simply "Faculty of Natural Sciences,"
"Faculty of Literatures," "Faculty of Politics and Economics," and so forth). The campus revealed an attempt to com-
bine Western design with traditional Islamic architecture.

When Cairo University opened its doors in December 1908 its founders intended it to become a haven for liberal thought. The ruling British administrators suspected that "liberal thought" meant the development of a distinct Egyptian identity, a threat to their interests.

By the time Kamal enrolled, the campus seemed weary from overuse and neglect. Classrooms hadn't been painted for years; the sound systems in the lecture rooms were barely intelligible. Kamal recalled sitting with students of all socio-economic backgrounds in musty, poorly ventilated lecture halls, the air outside yellow from diesel fumes, the light filtering weakly through windowpanes that hadn't been cleaned in years. The wealthy students found attending class demeaning. "For us it was a new, promising world," Kamal said about his early days as a student. "But some things were not as they should have been. There was moral decadence in the air. We felt alienated. We felt a responsibility to bring Islam into the classroom and campus life."

The future opening up to these young men should have glowed with promise. Under the tutelage of American and British secular modernizers, education, especially in the sciences and engineering, was considered the key to a future of comfort and progress—just it had been during the days of President Gamal Abdul Nasser and his socialist movement during the 1950s and 1960s. Bright young people by the tens of thousands had left their modest homes and farms, infused with the ideology of progress, drawn by images they'd seen on television and in movies and magazines. What they soon found, however, was a congested and decaying urban landscape, a political hierarchy marked by cynical cronyism and corruption, and an economic system that held out little promise to anyone outside the ruling elite. If there were ever crucibles for alienation and discontent in the Arab world, they were the coffeehouses and classrooms of Cairo University.

Kamal had been a devout Muslim before he matriculated. More than any other event, humiliation of the 1967 Arab military defeat at the hands of Israel had radicalized an entire generation of Arabs and Egyptians like Kamal. They came to view Islam not only as offering a spiritual direction to their lives but as an effective instrument of political action. In the 1940s, 1950s, and 1960s young Muslims, armed with secular Western ideologies like nationalism and socialism, had struggled against colonialism and backwardness. The 1967 *neksa*, "setback," molded the thinking and action of Kamal's generation; they rediscovered Islam with a vengeance. Before becoming a student at Cairo, therefore, Kamal had already made up his mind that Egypt needed to be set on the right path. Neither he nor most of his friends were from deeply religiously families; the opposite was the case. His generation's rebellion was as much as against the passivity of their parents, who had not dared to rock the boat of progress, as against the secular order. Kamal and others I talked to told me time and again that Egypt and its Arab sisters had lost the war against the tiny Jewish state because their parents' generation had ignored God's laws; Islam had become a stranger in its birthplace.

At Cairo, Kamal found the opportunity to preach to like-minded activists and to organize. "My heart and mind were made up," he recalled. He enrolled in courses in the social sciences and humanities and Islamic political thought. More important than the classes, however, were the meetings that took place outside of the classrooms, the long hours spent with activists praying, discussing politics, and reading radical religious texts. By the mid-1970s, Islamist students had become deeply entrenched; Kamal was among their ranks.

Conventional Western wisdom has it that a fundamental religious resurgence began to spread across Egyptian society in the 1970s and 1980s, somewhat similar to the way the Baptist revival movements swept across northern Europe and America in a different century. At bottom, it was a response to rampant industrialization. As millions of Egyptians, and millions more Arabs in other countries, found their village way of life overwhelmed by soulless progress and urban decay, they turned to the wisdom of the Prophet for the kind of solace that the empty promises about quality of life being made by the state bureaucracy simply could not match. Conventional or not—true or not—this version of events bears no resemblance to how Kamal explains what happened.

To Kamal's generation, religion was not some means to an end; it was an end in and of itself. It was not abstract but concrete. Al-shabab began changing the world by changing their world: campus life, they began to insist, had to reflect the teachings of the Prophet. There would be no more mixing among male and female students; Western dress would be replaced by the traditional white tunic. Men let their beards grow thick and untamed. Kamal and his friends didn't rely merely on moral persuasion. Recalling his own youthful exuberance, Kamal admitted, without the slightest hint of regret, that he and his cohorts had enforced Islamic rules. "Occasionally al-shabab would rough up male students who did not heed warnings against fraternizing with their females.

"We felt empowered and radicalized by our student experience. We dominated student unions and ended the hegemony enjoyed by the atheists and secular nationalists for over three decades. We established authentic Islamic communities and Islamic administrations to run our affairs. We provided social services to poor students who could not afford to purchase reading materials or pay for transportation and food. We ran summer training camps where students received religious sermons and physical [i.e., paramilitary] exercises. Some of us thought if we could Islamize universities, in a few years we could also Islamize government and society."

Of the dozens of 70s-era activists I spoke to, nearly all regularly cautioned me against the Western tendency to explain the rise of Islamism in purely socioeconomic and political terms. Such explanations, they felt, distorted and trivialized what their movement was all about: a spiritual and moral quest to halt, *not* merely to moderate, the secularization of society. It could not be understood as "a developmental crisis." "We did not sacrifice the flower of our youth, the best years of our lives, in prisons to get jobs and earthly rewards. Our aim is to please God. The West cannot comprehend our spirituality and religiosity as long as it is blinded by materialism."

The depth of Islamic spirituality, the extent of its reach into daily life, would be hard for even fundamentalist Christians to comprehend. For most Christians, prayer is either deeply private or something that happens on Sundays, but even those who pray every day are rarely regarded as fanatics. Fundamentalist Christians in the United States feel deeply and strongly about some issues—abortion and school prayer—but most want to accommodate those changes within the context of a secular state and material comforts. Most Christians would not call for the separation of the sexes, for example. For Islamists like the young Kamal, spiritual life and daily living were even more closely intertwined; it was the emptiness of secular progress represented by Western prosperity that set them on their spiritual quest.

No one was more important to that quest than Sayyid Qutb, an Egyptian poet turned Islamic radical who had been hanged by the Nasser regime in 1966 for alleged subversive activities. As early as the 1950s, Qutb sowed the seeds of jihadism in Egypt and throughout the Muslim world. He was a member of the Muslim Brotherhood after it had been radicalized by the 1949 assassination of its founder. Indeed he was the first contemporary thinker to define "jihad" not in terms of a conflict focused on a specific target in a specific era, but as an "eternal revolution" against any and all enemies, internal or external, who had usurped God's sovereignty.

Kamal's generation revered Qutb as the spiritual father of their movement. They referred to him as a *shahid*, "martyr," the highest praise that can be bestowed. "Qutb showed us the way forward," Kamal told me, gesturing forcefully with his hands.

As a young man, Qutb was a promising writer who composed belletristic poems and essays. In 1948 the Egyptian ministry of education sent him to the United States, where he took classes at Wilson's Teachers' College in Washington, D.C. He earned a master's degree in education at the Colorado State College of Education (now the University of Northern Colorado), and afterward started coursework at Stanford. The postwar Egyptian nation builders had supposed that sending their best and brightest young minds to study in Britain and America would give them the tools to establish modernity back home. Far from discovering cultural enlightenment, Qutb was repelled by the obsessive materialism and what he perceived as the spiritual barbarity of "the American man." On his return to Egypt, he published a scathing attack on the American way of life in the magazine *Al-Risala* titled "The America I Have Seen." This essay, later published as a book, colored the perceptions of America for generations of Islamists. Americans, Qutb wrote, worshiped at the altars of applied science and productivity, while shunning universal truths and human values:

This great America: What is its worth in the scale of human values? And what does it add to the moral account of humanity? And, by the journey's end, what will its

contribution be? I fear that a balance may not exist be-
tween America's material greatness and the quality of its
people. And I fear that the wheel of life will have turned
and the book of time will have closed and America will
have added nothing, or next to nothing, to the account of
morals that distinguishes man from object, and indeed,
mankind from animals.

As Qutb made his views public in newspapers, magazines, and speeches, the government accused him of being a
leader of an underground paramilitary wing of the Muslim
Brotherhood. Charged with trying to assassinate President
Nasser in October 1954, Qutb was thrown in prison and tortured, driving him deeper into militancy. How, he asked his
followers, could Muslims—Nasser's security services—inflict
suffering on believers like him? Before his execution in 1966,
Qutb managed to smuggle out, chapter by chapter, the manuscript of his last book. *Milestones* called for the creation of
hakimi, a "Godly society," and summoned a rising generation of believers to action. The call to arms was answered by
men like Kamal and Al Qaeda's Zawahiri, who in 1967 established the first jihadist cell in the Arab world. Qutb envi-
sioned them pointing the way toward the ultimate goal: to
purge Muslim society and politics of *jahiliya* and restoring
hakimiya to earth. As Qutb wrote in the introduction, "I have
written *Milestones* for this vanguard, which I consider to be
a reality about to materialize."

The martyred poet could hardly have imagined the influence his work would have. More than anyone else, Qutb

had shown him the way forward, Kamal told me. He and other young idealists were moved as much by Qutb's dignity and courage under torture as by his words. Zawahiri wrote in his own memoir, called *Knights Under the Prophet's Banner* and released after September 11, that Qutb's words acquired a deeper resonance because of his refusal to appeal to President Nasser to spare his life. Jihadists have dreamed of following in Qutb's footsteps on the road to martyrdom.

Kamal, the promising young student of political science, found himself on the same path. Along with hundreds of others—including Zawahiri—Kamal was arrested in 1981, immediately after the assassination of Sadat, and accused of inspiring those who did the shooting. Undercover agents took him to prison, where the torture sessions began immediately. A member of the Muslim Brotherhood recalls seeing Kamal's bloodied and swollen face the morning following his arrest.

Though they may not have realized it, the Egyptian government had in effect instituted a course in radical politics within the prison system. Within a few months Kamal had galvanized scores of fellow inmates, some interned for their militancy, others for political agitation, still others for ordinary crimes. Building on his student activist experience, Kamal inspired the other prisoners, so much so that they soon called him "emir." It was a pivotal moment in his life. When I spoke to his former fellow inmates, two words surfaced in their recollections of Kamal—*shuja'*, "courage," and *qiyadi*, "charisma." They all saw him as one of the most vital voices of the Islamic movement. Emir Kamal found himself in a position to shape the intra-jihadist debates that would play out over the next quarter century.

Yet in all of our conversations Kamal steadfastly refused to refer to himself as a leader. He always kept his focus on his religious mission, stressing the doctrinal and ideological harmony among those now calling themselves jihadists: "The ties that bind us are ingrained in our religious beliefs and an eternal quest to live a full Islamic life. I mean Islam as a complete way of life. Forget about personalities; we are God's instruments and a community of believers. We are all foot soldiers in this epic struggle. There are no hierarchies or class differences that separate us from one another. We are an egalitarian community." (Egalitarian or not, the jihadist movement's leadership emerged from among the social and political elite. Bin Laden and Zawahiri, who has a medical degree and came from a prosperous family, are cases in point. In contrast, in Egyptian terms, Kamal came from the middle class.)

We were sitting on the terrace of a coffeehouse on a little side street one January evening. Children were playing next to us. Other patrons were smoking and drinking thick, sweet Egyptian tea. His response struck me as a bit disingenuous. Ideologues of all stripes, messianic and secular, have often insisted they were claiming power not for themselves but for their cause. Nonetheless, rather than press him further about his own role, I asked him to explain the part played by leaders such as Sayyid Qutb and Sheikh Abdullah Azzam, spiritual mentor to bin Laden, and by the Afghan Arabs who fought the Soviet occupiers in the 1980s. These were charismatic leaders whose words fueled immediate action. Kamal refused to be drawn into my terms. Islamist leaders had never been crucial to the movement, he insisted, nor had they sought out the spotlight. "Do not view us through Western lenses," he cautioned me. "Let me repeat once more: We were not in it for money, nor for personal or political gain. Had we overthrown the Egyptian government, we would have surrendered power to an *ulema* ["religious scholars"] council to rule."

Instead, figures like Qutb stood as symbols of suffering. "When I was in prison, I knew that there were many people whom I had never met, who were familiar with my story and saw me as a symbol of resistance and defiance. They considered me an example to be imitated. For many, particularly al-shabab, jihadism was the visible expression of their own rage against oppression and injustice. Witnessing the political repression of well-known militants created within them a feeling of hopelessness, and along with that a desire for revenge. I felt all those vibrations around me, that I was living in a society that could offer thousands of martyrs ready to oppose injustice. It was an historic moment!" Kamal's voice had become emotional, and I could hear the tones that had inspired so many over the years.

Though he warned me against interpreting the Islamic movement in sociopolitical terms, proto-Marxist words such as "oppression," "injustice" and "hopelessness" permeated Kamal's speech. I wondered what he meant by them. "Our very identity, religion, and culture are threatened by corrupt secular influences. Muslim rulers are figureheads beholden to their Western masters. They oppress their citizens, while they cow before foreign enemies. The Arab state is sovereign in name only; it is penetrated and defeated by outside powers. Why are you shocked that we resent and abhor the existing unjust order at home and abroad?"

The jihadists I met in Cairo had all been suckled on a rhetoric of persecution and suffering, which is why martyrdom remains their highest expression of admiration. They shared a siege mentality and were driven by a desire for revenge. The Islamist lawyer Montasser al-Zayat provides examples of this phenomenon. In the early 1980s, along with Kamal and Zawahiri, Zayat served time in prison for involvement in the Sadat assassination. Since then, he has become one of the leading defenders of jihadists in Egyptian courtrooms. He has published two memoirs *Islamic Groups: A View from Within,* which was serialized in the Arabic-language *Al Hayat,* and *Ayman al-Zawahiri as I Knew Him.* Both books—which I will explore in greater depth later— reveal in excruciatingly gruesome detail the persecution that forged the bonds of solidarity between the jihadists. The prisons and torture chambers proved effective incubators for generations of activists.

⟨⟩

The radical tide swept well beyond the campus of Cairo University and joined with external militant forces. New jihadist cells were being formed, marking the beginning of the next phase in what was becoming a life-or-death struggle against the secular order. Those forces had long existed in Egyptian society, but they had always been marginal. By the late 1970s, however, the militants had tasted power. Again, however, Kamal insisted that religion remained the focus. "Some of us dreamed of capturing the state and instituting Islam. It would stand as an alternative to communism and Nasserism."

Why, I asked him, had the movement had gone underground? Why had his generation not sought a political solution, in the way that the Muslim Brotherhood had. I knew that the issue of openness was not a straightforward one. In many Mediterranean cultures, the relation between what is hidden and what is apparent, between public declaration and subterranean intrigue, stretches back in time. That which may be portrayed and spoken of directly against that which can only be known through the hand of God, indeed the very distinction between Western, secular, "objective" knowledge and a deeper spiritual truth, touches the foundations of social life across Muslim societies and was for centuries a leitmotif of Ottoman art and literature. Today these same tensions influence the way state power is organized. Understanding Kamal and his generation—grasping their dreams and their strategies—requires being mindful of that heritage. What lies beneath erupts in periods of crisis because it will not become integrated into any man-made system. This very same dynamic was at work in the relationship between communism, Nasserism—a form of Egyptian nationalism that formed around this charismatic president—and political Islam.

To Kamal, communism and Nasserism were contending anti-faiths, secular expressions of the palpable force of evil that, like Islam, had been suppressed by the European colonial powers after the fall of the Ottoman Empire. Both, he says, were hibernating, waiting for the right conditions. The rapid decline of the British Empire after World War II provided the historic moment. "They [communism and state nationalism under Nasser] dominated Arab politics since independence in the mid-1950s. They infiltrated state bureaucracies and institutions and excluded Islam. It has been a long, bitter struggle between atheism and secularism, on one side, and Islam and *iman*, 'belief,' on the other. We wanted to bury the former for good."

It is a Manichaean view of life—believing that human history is shaped by a titanic struggle between absolutes of good and evil. There is Islam and the Islamic way of living, and there is Satan, ever-present evil that is forming cells of corruption and debauchery in the form of democratic, secular politics. For a Godly life to be possible, its enemy must be annihilated. Ironically, it was the program of secular development—free public education combined with dramatically expanded access to university and technical training—linked to the language, if not the reality, of citizens' rights, that enabled these bright young Islamic activists to find each other and to use their relatively expanded freedoms to engage with the broader movements then energizing the Arab world.

If there was a powder keg in that volatile geography—a place where the Manichaean forces might confront each other—it was Lebanon. In the mid-1970s a civil war erupted there between Christians and Muslims that played a vital role in radicalizing Kamal. Lebanese Christian militants were pioneers in putting religion at the service of politics, consciously drawing parallels between their actions and the Crusades. They often celebrated their conquests of Muslim neighborhoods, their chests adorned with giant crosses. Kamal and his coterie of fellow political science students felt they could not sit quietly by while, for the first time in what seemed like centuries, Muslims appeared to be rising up in their own defense. When a couple of years later bands of Islamic students in Iran turned their dictatorial shah out of power and humiliated the United States, it again seemed inconceivable for any committed student of conscience to step aside while God and history were on the march. "Al-shabab interacted with and responded to both developments and longed to take action," was the understated way Kamal expressed it.

Kamal's words revealed the scope of the political struggles taking place within the Arab world in the 1970s. On the one hand, radicals like Kamal had already formed a powerful social base. He had even started to think about the unthinkable—seizing power by force. At the same time, the old order of secular nationalists and socialists had fallen into smaller and sharper splinters, divided by irreconcilable personal and ideological differences. Even the pro-Western Anwar Sadat was willing to ally himself with the "devil," his critics charged, to defeat like-minded rivals, Nasserites, and leftists. Sadat wagered that he could keep Islamists under control and use them as deterrents against his secular enemies. Little did he know that Kamal and his generation would prove implacable: "We were not for sale, we would not compromise our Islamic values. Our generation was independent and rejectionist and in a hurry to substitute the word of God for man's."

⊙━━✦━━⊙

As became clear from my conversations with Kamal and others, the birth of modern jihadism in Egypt—and therefore in the Arab world as a whole—involved the political stories of Gamal Abdul-Nasser and Anwar Sadat, in whose assassination Kamal was accused of having played a key part. If Nasser had seen himself as a kind of Egyptian Ataturk—the Turkish leader who brought that country into alignment with the West—replacing a traditional society with the discipline of the modern state, Sadat had engaged a more complex strategy. Recognizing Islam's profound roots in Egypt and keenly aware of the disaffection brewing within Kamal's generation, Sadat sought to create an Islamic counterweight to the nationalist and socialist forces that had dominated Arab political life since Egypt's independence from England, particularly after the Suez Crisis in 1956, when Nasser nationalized the canal, and the British and French were forced to back down.

Sadat wanted to escape Nasser's shadow. Nasser had delivered stirring speeches, capturing the hearts of millions of Egyptians and Arabs, but he hadn't been able to deliver much else except bloated bureaucracy, entrenched authoritarianism, and expansion of Soviet influence in his country. Sadat saw the direction the Cold War was headed and pulled Egypt away from the Russian and into the American camp. It was perhaps the right move to make, but it carried an incalculable risk: If he wanted the aid, arms, and security training that America offered, he had no choice but to make peace with enemy number one of the Arab world, Israel. The Islamists certainly bore no great love for the socialist model, but by ending Egypt's experiment with socialism, Sadat was also giving the country over to the soulless market forces that Sayyid Qutb had so vehemently denounced. It couldn't have worked out better for the young Islamists, indirect beneficiaries of Sadat's drive to build his power base and create his own legacy. Though they might have taken advantage of the opportunity, Kamal insisted, they were never the pawns that Sadat played them for, and they never sacrificed their integrity.

Still, I asked Kamal, couldn't we say that Sadat, the great peacemaker, winner of the Nobel Peace Prize—a Western award, admittedly—was most responsible for strengthening the Islamist movement? "No, Sadat did not propel Islamists to the top of the popularity polls," Kamal retorted. "He did not bring about the Islamic awakening that swept through Egypt and the rest of the Muslim world. All he did was to allow the Islamists to compete on an equal footing with the communists and Nasserites."

Sadat's gestures to political Islam were by no means trivial. In the early 1970s he amended the Egyptian constitution, enshrining Islam as "*the* source," not just "a source," of legislation. But despite his initial flirtations with the Islamist movement and his attempt to cast himself in the role of the "pious president," Sadat gradually lost favor with the Islamists. "He never intended to implement the shariah," Kamal said dismissively. "He revealed his ugly face by not delivering on his promise." As support waned and the opposition became more vocal, Sadat responded with an iron fist. First he locked up the Islamist leaders; then he cracked down on their rank and file.

Eager as he was to distance himself from Sadat's Islamist overtures, Kamal paused in his attack on the slain president. "Still," he added with a faint smile, "thanks to Sadat, a new healthy Islamic generation was born unfettered and unscarred by persecution and torture like the previous generation in the 1950s and 1960s. I was never questioned or harassed by security forces. My generation had no complexes and no nightmares and we have paved the way for all subsequent Islamic and jihadist waves in the Muslim world, not just Egypt." Then, anxious that I not construe him as being in any way grateful to Sadat, Kamal returned abruptly to the role played by socialization, upbringing, and popular culture in fueling Islamic activism: "We were breastfed on Islamic values that shaped our existence and conduct. You have to understand the powerful longing to make Islam supreme, which was the driving force behind our awakening in the 1970s. Sadat was just a catalyst enabling us to express our sentiments publicly and organize ourselves."

Still, it seemed to me that Sadat had provided the opening to people like Kamal, many of whom had suffered—or, like Qutb, been martyred—for their religious convictions. "Why," I asked Kamal, "did you risk it all to take arms against the very man who had indulged you? Why bite the hand that fed you, or at least did not torture you?" Though I am fully at home in the Muslim world, my question seemed to symbolize to Kamal how far removed from its realities I was. "You must understand," he began, speaking as patient teacher might to a dull-witted student, "Islam is a complete way of life. It encompasses all personal, social, and political aspects. There is an organic link between Qur'anic law, shariah, and political authority. Although, on the whole, Muslim societies live by Qur'anic law, political power applies secular rules. There will be no security as long as political authority is not based on God's sovereignty. There will be no peace."

"There will be no peace." His words portended so much of what was to follow. But at the time of our discussions my interest remained focused on Sadat's assassination. I still couldn't understand why Kamal and his generation had risked everything to kill him. My stubborn pursuit angered him. He ran down a long shopping list of grievances. At the top of the indictment was Sadat's single most unforgivable offense: He had reneged on his pledge to make shariah the law of the land. Instead Sadat insulted the community of religious scholars and issued decrees restricting criticism of his policies. One dictated that insulting the president would be punishable with imprisonment.

But it was of course the 1978 Camp David treaty with Israel that proved to be the last straw. Sadat "shocked the sensibilities and beliefs of the Egyptian people" and enraged activists like Kamal. There were no subtleties in this equation. "Sadat stabbed us all in the back." Kamal's position on Israel was simple. In his 1986 prison essay, he argued that one of the fundamental issues facing the ummah worldwide was "the existence of Israel in the heart of the Muslim world." Muslims ought to deal with Israel in the same way that European settlers in America had dealt with Native Americans. It seemed a troubling and ironic comparison, one that would hardly create sympathy for the cause, but of course it wasn't written for Western consumption. He had meant to convey a sense of inevitability, that it was essentially the manifest destiny of the Muslims to take back Palestine. The struggle between Muslims and Jews "will not end except when one camp excises the other, and both sides must recognize this fact."

Perhaps not surprisingly, Kamal's views on Israel had evolved by the time he and I talked. He no longer believed that Israel should be destroyed. However, he insisted that it end its oppression of Arabs and cease the occupation of their lands. To survive, Israel must shed Western legacies and seek acceptance by Muslims.

While his views on Israel might have tempered, Kamal's opinions of Sadat had not. Then and now, Sadat represented something that could not be tolerated. His betrayals were manifold, ranging from harassing the Islamist movement to embracing the infidel by allying Egypt with the United States to allowing commercialism to corrupt Egyptian society. It was the first and the last—domestic politics rather than foreign policy—that had fueled Kamal's generation's revolt. Had Sadat simply implemented Islamic law, as he had promised he would do, jihadists would have turned a blind eye to his opening up Egypt to the United States, even to his signing "the shameful peace with the Jews." Sadat represented the greatest of all dangers, the one from within.

Kamal and others I spoke with made it clear that part of their rage against Sadat was focussed on his wife. Jihan Sadat, they believed, had desecrated Islam's moral directives. Kamal thought that she had convinced her husband to pass secular legislation and to empower women, violating the most fundamental codes of Islamic morality. Islamists often cite the televised images of Jihan dancing with President Jimmy Carter at a White House reception as a defining moment. Whether or not Sadat himself was shamed by her wanton display, Islamists were. Sadat's moral authority as leader of the Muslim community collapsed. When Sadat and Jihan personally offered refuge to the Shah of Iran after the Iranian Revolution in 1979, there was no turning back.

On October 6, 1981, Sadat, looking resplendent in a ceremonial uniform, was on a reviewing stand, watching a military parade marking the annual celebration of the 1973 war against Israel. The entire Egyptian political apparatus was assembled, also in dress uniform. With the crowd distracted by an aerial display of fighter jets, a military truck came to a halt in full view of the television cameras. Four men jumped out and began spraying the reviewing stand with automatic gunfire and throwing grenades. Sadat was killed during the ensuing panic. In a phrase that was to be repeated everywhere, the ringleader shouted, just after opening fire, "I am Khaled al-Islambuli. I have killed Pharaoh and I do not fear death."

I wanted to know how Kamal and the others had re-solved to assassinate Sadat. I knew that he had directly participated in the planning, though up until then in our conversations he had not admitted as much. It was a critical juncture. I knew that he might well cut me off, abruptly end-ing our conversations and my access to others in the move-ment. He inhaled deeply.

"The decision to assassinate Sadat was spontaneous and amateurish," he said. "It was not a well-coordinated opera-tion, and it succeeded by a miracle." Deeply religious Mus-lims like Kamal adhere to an austere form of belief, one that does not—strange as it might seem to outsiders—engage in elaborate rituals or turn to the "miraculous." Nonetheless, he used the word *miracle* to convey divine intervention on behalf of Sadat's assassins. They had been short on ammu-nition and had no inside connections with Sadat's security detail. Al-shabab had "made history" due to the ineptitude of the Egyptian bureaucracy. Lest this seem too sarcastic a way of putting it, Kamal quickly added that the perpetrators knew very well that they would not survive this daring at-tack and were "guided by a spirit of martyrdom.... They aimed," as he put it, "at pleasing God." That spirit of martyr-dom is what linked jihadists of all generations together.

⊙━◆━⊙

Over the course of our conversations, Kamal and I met in a number of locations—his office, at a small side street café, the campus of Cairo University. We often met on Fridays, the day of rest in Muslim countries, late in the evening, after he had prayed and had dinner with his family. Sometimes he arrived late for our interviews. He would apologize and explain how hard it was to leave his wife and children alone at home on a Friday. He spoke proudly of his children and about how they were living an "Islamic life" uncorrupted by "imported television values." His biggest challenge, he admitted, was simply getting by and paying for things—heat, food, shirts, shoes, school notebooks. As a leading Islamist, the taint of an assassin's blood on his hands, he was all but unemployable. His close relatives—and even his children— had been blacklisted by the authorities and were unable to hold jobs in the public sector. "They are punishing my family for my beliefs," he once said, shrugging.

But Midan al-Tahrir, "Liberation Square," was one of his favorite places to meet. Located in the heart of Cairo, it is congested with traffic; even at night an unending stream of exhaust-spewing old cars chug past. We would sit near the entrance of the subway on a cement block. I suspected security weighed heavily in his choice of where to meet. Kamal believed that he was under constant surveillance—which he probably was—and he wanted to show that he had nothing to hide. We met there one particular night during Ramadan. An enormous neon Pepsi-Cola billboard loomed over us. A group of children had taken over a sliver of the square to play soccer. Others were walking back and forth selling gum and barbecued corn. Not far away a young couple sat uncomfortably on a concrete bench; the girl was wearing a scarf and the young man stared intently into her eyes. I thought it a little ironic that here I was meeting with someone who, if he could, would forever ban open interaction between men and women and forbid such shameless flirting.

I asked Kamal how women would fare in an Egypt ruled by Islamic law. "The state does not possess the right to legislate or alter holy texts under any condition," he responded with some heat. "No one can dictate that there should be equal inheritance rights. Women must not be permitted, like men, to marry more than one husband at a time. I can accept a woman judge as long as her authority is limited to women. I reject *wilayet al-mar'ah* ["state ruled by a woman"] because women are not psychologically equipped to lead men. An Islamic state must fully implement Qur'anic laws and refrain from any legislation on these laws."

Kamal's response expressed the sentiments of most Islamists, who view women's rights as a form of Western encroachment and imperialism. One of Kamal's associates, now an attorney, simply told me to shut up when I pressed him on why activists seemed so terrified by female empowerment: "I am a son of the Islamist movement. I will do everything in my power to prevent women from becoming judges. They belong at home with the children."

I reminded Kamal that his generation was rethinking its old views and developing a new political blueprint, one that was edging toward democracy. And democracy, surely, required individual rights and equality under the law? He wouldn't be budged. Equality for women ran up against an absolute. "I value communal rights more than individual rights. In an Islamic state, the individual is not free to do what he wishes. There are limits ordained by God's laws, which supersede any human authority." Here was a reminder that despite the ways in which Kamal and his generation have modified their youthful views, their ideas still clash with Western liberal democracy. Islamic democracy will not be a carbon copy of Western liberal democracy; it will be deeply rooted and colored by local traditions and values.

⚜

From the moment I met Kamal, I had been struck by the notion of an entire generation of college students, inspired by the rhetoric of Islamist ideology, politically polarized, ready to do battle to bring God's kingdom to earth. Impatient and uncompromising, they had believed fervently in the forced Islamization of state and society. Kamal acknowledged that he and his generation had made a fatal mistake. Rather than trying to build support from the bottom up, they became fixated on capturing political power and imposing their religious order from the top down. Instead of galvanizing the masses, as they'd imagined, their methods—like the random killings of tourists and ordinary citizens—repelled most Egyptians. God-fearing and peace-loving by nature, the Egyptian public withdrew any sympathy they might have shared with the revolutionaries.

"We were naive, arrogant, and immature, fired up by the spirit of youth," Kamal told me in one of our last conversations. It was early in the evening and we had gone to a political rally organized by a moderate Islamic party, the kind of gathering Kamal wouldn't have dreamed of attending thirty years ago. He had been impressed by the proceedings and had listened to the speeches attentively. We now sat on a bench facing the Nile. I do not know if it was the effects of the place or the fact that I'd become something of a fixture to him, but Kamal turned more self-critical than I had ever seen him. He suddenly seemed drained. "We had big dreams but few resources, and there was a pronounced gap between the means at our disposal and our ambitions. Gradually, we lost sight of the balance between ends and means and fell into the trap of armed escalation with the government. We were no match for its powers."

Every movement goes through its phases, the jihadist movement no less than any other. "We have come a long way since the 1970s, even though our journey has been painful and costly. We had to learn by trial and error. We were fortunate to be the pioneers. But we had no support network of wise men and spiritual mentors to guide us through the minefields. We were on our own, struggling against great odds and challenges."

"What if you had succeeded, Kamal?" I asked. "Were you truly prepared to establish a viable Islamic government?"

"Thank God, we did not win, because we would have constructed a state along the same authoritarian lines as the ones existing in the Muslim world. We had no vision or an intellectual framework of what a state is or how it functions and how it should be administered, except that it should express and approximate the Islamic ideal. While I cannot predict that our state would have been totalitarian, we had little awareness of the challenges that needed to be overcome. We were terribly naive and did not appreciate the complexities of society and the requirements of social and political change. Jihadist organizations, then and now, failed to build a broad social base of support. We only became conscious of what was required during our prison years."

The honesty of his reply left me dumbfounded. He was suggesting that he and his generation of jihadists had been more an elitist vanguard than a mass movement. This contradicted their claim that they had represented the ummah's popular will. "Why," I asked him, "did not the people come to your aid instead of turning against you?"

"It is not true that the people turned against us," Kamal shot back. "They were passive and did not join the fight on our side. The masses hate authority. My generation and the subsequent one simply did not know how to mobilize at the street level. At first the people looked to us to transform society and governance. They viewed us in a purer light than the secular corrupt rulers. In the 1970s and 1980s jihadists and Islamists became a power to be reckoned with in society. But they did not know how to translate popularity into real influence. They relied too much on muscle and not enough on political persuasion and mobilization."

His tone became that of the mellowing but still unrepentant revolutionary who had missed his moment and was prepared to look back on it with a measure of sentimental self-criticism. I asked whether a revolution in the name of God could be the same as one launched in the name of material things, such as food or shelter or land. Wouldn't their revolution have had to been one in which shariah would get imposed by fiat, as it was and is in Iran? Isn't jihad by its very nature revolutionary rather than peaceful? I was consciously trying to provoke him.

"Listen," he said, his voice rising, "we did not fire the first round in this battle. The rulers had closed all avenues for a peaceful transfer of power. We had no choice but to take up arms to raise Islam's banner. The fight was imposed on us." No matter how cool and analytical Kamal tried to be when articulating the struggle of his generation, he remained haunted by the prison years. The first time the subject was broached, it all came pouring out—a rush of memories and feelings.

"It is true that we began our ideological revisions while in prisons in the 1980s. But some of us arrived at different conclusions as to what could and should be done to resist aggression against our Islamic identity. The prison years also radicalized al-shabab and set them on another violent journey. The torture left deep physical and psychological scars on jihadists and fueled their thirst for vengeance. Look at my hands—still spotted with the scars from cigarette burns nineteen years later. For days on end we were brutalized—our faces bloodied, our bodies broken with electrical shocks and other devices. The torturers aimed at breaking our souls and brainwashing us. They wanted to humiliate us and force us to betray the closest members of our cells.

"I spent sleepless nights listening to the screams of young men echoing from the torture chambers. A degrading, dehumanizing experience. I cannot convey to you the rage felt by al-shabab who were tortured after Sadat's assassination.

Some left the prisons and the country determined to exact revenge on their tormentors and torturers. The authorities' brutal methods nourished fanaticism and sowed the seeds for more violence and bloodshed."

Such did not seem the case with Kamal. Prison had drained the violence out of him. By the time of our last conversation in Liberation Square, he seemed dispirited. He felt wasted in Egypt. *Al-Shaab* had been shut down by the authorities. Incredibly, he spoke of enrolling in an American doctoral program, even asking me if I could help him obtain a scholarship in Islamic studies at an American university. His interest appeared genuine; he told me that his dream was to complete his higher education. He ignored the glaring irony of the request, though it represents an outlook shared by so many radical Islamists: a fundamental antipathy to what America stands for and yet a clear appreciation for what it can offer.

❦

Of all the militants I have come to know over the years, Kamal stands apart. Privy to the innermost councils of the radical Islamist movement for more than a quarter century, he is perhaps the most open and candid of them all. I have no doubt that he was fully aware of the tragic developments in the years that followed our first conversations, and he likely played a role in deciding whether they should go forward or be scrubbed. To this day he remains a keen observer of activities, even those by the generation that succeeded him, the one that rose in the 1990s and carried out the September 11 attacks.

Kamal's story also reveals the changes that have taken place within jihadism over the decades. Once a proponent of armed resistance against pro-Western secular Muslim rulers, he is now struggling to reconcile himself to nonviolent political action. He has taken the lead in trying to chart a new course for those who are fed up with killing and getting killed but are still wedded to the dream of an Islamic state.

Kamal continues to attract a large following among the youngest generation of radicals in Egypt. Tens of thousands of their fellow jihadists are behind bars and many more face a grim future. The choices they face are painful. Should they answer the call to war against the United States, the call that echoed down from the mountains and valleys of Afghanistan and is now coming from Iraq? Should they join the World Islamic Front for Jihad Against Jews and Crusaders, created with such fanfare by Osama bin Laden and their countryman Ayman Zawahiri in 1998? Or should they learn from Kamal and his generation and embark on a nonviolent journey?

When I first got to know him, Kamal was offering a way dramatically different from global jihad. Although he did not command armed brigades and suicide bombers, as do bin Laden and Zawahiri, his authority remains beyond question. Scores of activists throughout the Muslim world told me that Kamal's writings and manifestos had influenced their fight against secular rulers.

But his jihadist journey seemed to have come to an end. He and others of his generation had been destroyed by their struggle: thousands perished and thousands more were still rotting in prison cells. "What do you tell the families of the martyrs, and how would you take care of their beloved ones?" he once cried. This from a former jihad emir who almost certainly issued orders that led to the deaths of those who obeyed them.

"My release on October 16, 1991, was a miracle," he once admitted to me. "I cherish every day in freedom and consider it a gift from God, particularly since the authorities occasionally rearrest me. I want to be useful and make the best of my life."

II

THINGS FALL APART

IN THE MIDDLE OF THE 1970S the center began to fall out of the Arab world, and when it did it nearly destroyed Lebanon, my homeland, shattering the long and peaceful coexistence between Christians and Muslims. Lebanon descended into full-scale sectarian strife that would last for fifteen years and bleed the country dry. The Lebanese Civil War, as it is now called, was an internal struggle for the soul and future of the Arab and Muslim world, and it shook the very foundation of Arab society. Its reverberations were felt far beyond tiny Lebanon's borders.

In our conversations, Kamal and his confreres impressed on me how what happened in Lebanon drove them to extremism: "The ummah did nothing while Muslims were being persecuted by the crusaders and the Zionists," Kamal said. He and his cohorts monitored the civil war on radio, on television, and in print media. "Al-shabab were eager to join the fight on behalf of our Lebanese and Palestinian brothers battered by Israel's military and its local agents," he said. "We were moved to tears by their plight. But we knew that it was an impossible mission. We also recognized that the fight in Egypt had to take precedence over everything else. The outcome of the struggle in Egypt would determine the future of the Arab and Muslim world."

The civil war in Lebanon polarized a generation of Muslims and Christians, many of whom had been raised in tolerant conditions. They started bringing religion into their politics; there was no turning back now, they felt. As Assaad Chaftari, then a commander of the Phalangists, an extremist Christian militia, and later the equally hard-line Lebanese Forces, publicly admitted, "My comrades in the Phalangists and I thought that it was our duty to defend Christians and considered Lebanon the bulwark of Christians in the Middle East. We lived in a sea of Muslims. We thought that if we gave in to Muslim demands for equal representation in politics, it would be the end of the Christian population in the region."

Hicham Shihab grew up in a liberal, lower-middle-class Sunni home in Beirut, but joined al-Jama'a al-Islamiya in 1975, when he was thirteen, becoming an active militiaman. The rise of Christian extremism, he said, led him to discover religion and become an Islamist: "My friends and I felt that since the 1920s our leaders had been experimenting with bankrupt ideologies like Arab nationalism and socialism, which failed to liberate Palestine and restore our dignity as

Arabs and Muslims. We thought political Islam was the only means to undo the wrongs. We also believed that those Western ideologies were merely ploys to divert Muslims from their noble goals. Our preachers and clerics often told us that Arabs would regain their glory only if they reclaimed Islam and established shariah. Lebanon, we thought, should not be the monopoly of Christians."

Like others of his generation, Hicham looks back on those days with a wisdom that was hard-won. "We were very young and immature. We were children of the war. Our fanaticism stemmed from ignorance. Now we know better," he said with a measure of sadness. Today he is married and the father of four. He attends a Catholic university north of Beirut, works as a journalist and translator, and participates in Christian-Muslim reconciliation activities.

If the seeds of religious militancy were sown in Lebanon's fertile soil, their offshoots quickly reached into neighboring Arab countries. The Islamists I interviewed—even those, like Hicham, who had grown more tolerant—had been enraged by the sectarian strife in Lebanon, particularly when viewed in combination with the military and financial support given to the Christians by America, Israel, and the pro-Western Shah of Iran. They felt Muslim rulers had failed their coreligionists in Lebanon. As one Egyptian jihadist noted, "The predicament of our Muslim brethren in Lebanon reminded us of the need for a new Islamic army to defend the ummah and to replace pro-Western, secular Arab regimes with God's law." Hicham testified to the link between the

Lebanese civil war and the rise of Islamist militancy in neighboring Arab states. "I remember that our Islamist cell numbered more than a hundred and fifty young men, just from a small district in Beirut. But as soon as we started military operations, Muslim brethren from neighboring countries, especially Syria, snuck into Lebanon to help us fight the Phalangists. They believed that they were waging the highest form of jihad, a training stage for the global showdown with the infidels."

Perched on top of a hill, my village, Tal Abbas, was slowly entering the twentieth century when I lived there. The old stone houses were gradually being replaced by cement villa-style buildings. Nonetheless, narrow alleyways still connected the various neighborhoods and only the main street was paved. Villagers eked out a living from farming and depended upon money sent home by those who had emigrated to work in places like Boston, Montreal, New York, and Sydney. The village had long been a model of religious and ethnic tolerance. Surrounded by Muslim villages in the fertile Akkar valley in northern Lebanon, it was fully integrated with its neighbors. Muslims from nearby villages worked and traded extensively with us. My father, who hauled produce— vegetables, wheat, and citrus—to the Beirut market, employed several Muslim helpers with whom we often shared meals and celebrated family occasions. My brothers and I played volleyball with their children. Our elementary and primary schools included Muslim and Christian boys and girls. With a population of only three thousand, Tal Abbas was too small to have its own high school. We had no clinic, no public library, no telephone lines, no sports facilities.

That is not to say there were no distinctions between Christians and Muslims in Lebanon as a whole. Most of the doctors, lawyers, bankers, and army officers were Christian; most of the farmers, laborers, traders, and infantrymen were Muslim. Christians had dominated the economy and politics at the expense of the Muslims since the founding of the country in the 1920s; over time socioeconomic and political inequities widened and deepened the cultural-religious divide, while Muslims grew to be the majority. By 1975, social and economic resentment, coupled with disagreements over regional and international affairs, had split the Lebanese into warring camps along sectarian lines. But in Tal Abbas, Christians were not all wealthy professionals, nor were all Muslims simple laborers. Those differences did not exist: we were all poor. I don't remember feeling or sensing discrimination.

Akkar is a predominantly Sunni province, and though my family was Greek Orthodox, I learned the rewards and the challenges of living in the heart of the Muslim world. Muslims were willing to embrace and protect sizable communities of other faiths. Historically, Islamic societies have been diverse, open, and tolerant. Christian and Jewish communities survived and even thrived in the Muslim Middle East since the birth and expansion of Islam. Non-Muslims were never confined to ghettos, nor were they subjected to the brutal persecution their counterparts suffered in Europe. No forced conversions, no expulsions of the sort that followed the Catholic Reconquista of fifteenth- and sixteenth-century Spain ever befell Christians and Jews in the Islamic world. There were no Islamic pogroms. There were no Muslim Crusades. The Holocaust occurred in the heart of Christian Europe, not the House of Islam.

Since independence, Lebanon—whose population at founding was roughly equally divided between Muslims and Christians—served as a bridge between Islam and the Christian West. A compromise was found, whereby Lebanon would maintain both an Arab identity and a Western sensibility. The Lebanese model was symbolically important well beyond its borders, for it demonstrated that Western democracy and modernity were compatible with Islam, and that the two civilizations could coexist peacefully. Despite the political and social upheaval engulfing its Arab neighbors, Lebanon preserved, if precariously, a liberal, vibrant political culture and a laissez-faire economy. It became a playground for prosperous Europeans and Americans and a sanctuary for Arab writers, poets, and opposition figures.

Lebanon's open society, its free press, and its policy of sheltering dissidents represented a threat in a region ruled by strongmen. Dictators and autocratic kings reigned supreme in Syria, Egypt, Jordan, Iraq, Saudi Arabia, Libya, and the Sudan. It was difficult to keep track of the number of coups d'état in neighboring Syria until 1970, when Hafiz Assad seized power and established a military dictatorship that proved durable. The Gulf monarchies ruled by bloodline and religious mandate; the Baathists in Iraq terrorized their people into submission. Starting in the mid-1950s *mukhabarat*, or security services, from Egypt, Syria, Iraq, Libya, and elsewhere, as well as the CIA, the Mossad, and the Iranian secret police, began infiltrating Lebanon and chipping away at its democratic underpinnings by assassinating political opponents and provoking instability. In 1966 a leading liberal journalist, Kamel Mrouweh, who had dared criticize Nasser, was gunned down in Beirut. Ibrahim Qoleilat, a leader of a pro-Nasserite group, al-Murabitun, was implicated in the assassination. On April 10, 1973, a team of Israeli commandos raided Lebanon and assassinated three senior Palestinian leaders in their beds. Their funeral turned into one of the largest public gatherings in support of the PLO that Beirut had ever witnessed. Protesters loudly accused the government of complacency, even of collusion with the Israelis.

Lebanon's nightmare had begun. On February 27, 1975, a demonstration by local fishermen degenerated into violence. Several people were killed, among them a local Sunni politician, Ma'ruf Sa'd, leader of a Nasserite organization in Sidon. Left-wing parties accused the Lebanese government and the army of being behind the assassination. Bloody clashes broke out between the security forces and an alliance of leftists, pan-Arab nationalists, and Palestinians. This marked the first shot in the Lebanese Civil War. Soon Arab rulers were fighting one another through a proxy war in Lebanon, aligning themselves with various political and ideological factions and pouring gasoline on an already volatile situation. Beginning in the mid-1950s, Nasser lent a helping hand to like-minded Arab nationalists, while Syria armed and financed a paramilitary organization called the Syrian Socialist Nationalist Party. Not to be outdone, King Hussein of Jordan and the Shah of Iran backed the minority Maronite Christians, who were desperately scrambling to maintain the political privileges that they were given under the terms of an unwritten agreement made among Lebanon's ruling class at the time of independence. Ineluctably, inter-Arab rivalries overburdened Lebanon's fragile political system and exacerbated Christian-Muslim tensions, particularly after Sadat took Egypt out of the Arab camp and negotiated a peace treaty with Israel.

Equally important, Lebanon could not shield itself from the rising Arab-Israeli tension. Hundreds of thousands of Palestinian refugees had moved into Lebanon and developed a powerful political and paramilitary machine. After Jordan expelled the Palestine Liberation Organization in 1970, the PLO established a state within a state in Lebanon. The government was too weak to shoulder its security responsibility and force the PLO to respect the country's sovereignty. In addition, the Lebanese people were themselves divided over the PLO, which was seen as a legitimate liberation movement by a majority of citizens. When the PLO began launching attacks against Israel from southern Lebanon, Israel retaliated, striking Palestinian military and civilian targets in Beirut, Tyre, Sidon, and the Bekaa valley. Determined to seal its border and prevent further PLO infiltration, Israel focused on breaking the backbone of the Beirut government through punitive attacks against economic infrastructure, shelling

Beirut's international airport, electrical grids, and bridges. The Lebanese government—first run by President Charles Helou (1964–1970), a liberal Maronite, and then Suleiman Frangieh (1970–1976), a conservative Maronite—would have liked to close the border, but it had neither the public support nor the military muscle to do it.

In my village we were as divided as the rest of the country. Some young men joined pro-Palestinian leftist factions and others anti-Palestinian right-wing groups. The split did not run strictly along Christian-Muslim lines, though a majority of Christians vehemently opposed an armed Palestinian presence in Lebanon. Like my schoolmates, I was torn in my sentiments. My heart lay with the Palestinian refugees living in a nearby camp, Nahr al-Barid, where large families were stuffed into ramshackle hovels without water or electricity. But I also feared that my community would splinter further, and that my country would pay dearly in blood for the Palestinian armed struggle against Israeli occupation of their lands.

The Palestinian presence brought to a boil already simmering internal differences, socioeconomic inequities, and violent disagreements about foreign policy. Lebanon's prosperity had been driven by tourism, banking, and construction. But economic development had not trickled down to the vast majority of the population, particularly not to the rural Muslims who mainly worked in agriculture and manual labor. On the morning of April 13, 1975, a beautiful spring day, unidentified gunmen opened fire at a congregation outside a Maronite church in Ayn al-Rummaneh, a Christian residential neighborhood, killing four people. A few hours later, a bus was carrying thirty passengers, most of them Palestinians, through a Christian residential neighborhood, Ayn al-Rummaneh, on its way to a Palestinian refugee camp, Tal al-Za'tar, located in the heart of Christian East Beirut. In apparent retaliation, a band of armed Christian militiamen (from the al-Kataeb and National Liberal Parties) opened fire on the bus and killed twenty-nine people. The driver and one passenger, a Lebanese Muslim cleric, were the only survivors.

That massacre was the spark. For fifteen years, Lebanon became the killing fields of the Middle East. No family—Christian, Muslim, or Druze—escaped unscathed. I lost dozens of schoolmates and several family members, including my younger brother Bassam, then twenty-nine years old. Tit-for-tat sectarian killings and massacres terrorized the population and opened a gulf between the country's sects and religious communities. One day a right-wing Christian militia would kidnap and execute a dozen Muslims. The following day a Muslim-led militia would retaliate by shooting a dozen Christians. Roaming bands of militiamen prowled the neighborhoods, often at night, and snatched victims from their cars or the sidewalk.

Between 1975 and 1990 thirty thousand people were kidnapped. Most did not return. The families of these "disappeared" still have not come to terms with their losses. I have an aunt now in her eighties whose son, Elias, disappeared in August 1983, near Swayifat, next to Beirut's airport. Elias was in his twenties. To this day, every time I visit her my aunt tells me that "somebody, somewhere" has heard of or seen Elias, and that deep in her heart she knows that he is still struggling to escape from his kidnappers. I wish my aunt's tale were atypical, but it isn't.

One Saturday in 1975 the Phalangist militia kidnapped and killed hundreds of Muslims on the outskirts of Beirut, supposedly in retaliation for the brutal murder of several Phalangist members by Palestinian militiamen in Muslim West Beirut a few days before. The Phalangists, feared by Christians and Muslims alike, collected their victims, simple day laborers, as they left work at the Beirut port. According to the few survivors, some of the men were lined up against the wall and shot. They were the lucky ones. Others were tortured and stabbed to death; their mutilated corpses told of the horror of their ordeal. Some disappeared; some were dumped in dirt lots off the side of the road. "Black Saturday" shook the conscience of the nation, enraged the Muslim community, and instigated Muslim extremists into striking back harder at Christians.

The autumn of 1975 saw the beginning of what is now called ethnic and religious cleansing—particularly since the horrific events in the former Yugoslavia. The Phalangists attacked Tal al-Za'tar, a Palestinian refugee camp in the heart of Christian East Beirut, first ransacking, then leveling it. Palestinians survivors fled into predominantly Muslim West Beirut. In retaliation, Palestinian armed organizations, along with leftist and Muslim militias, raided Damour, a Christian town south of Beirut; they expelled its forty thousand inhabitants, looting their homes and shops. Tales of rape and killings spread more fear and rage.

～━◆━～

And on October 7, 1975 death came to Tal Abbas, my village. It was 3 A.M.—pitch dark. Some of the attackers came in pickup trucks, so that they could carry off everything possible—refrigerators, television sets, furniture. Those who arrived on foot planned to steal cars. For several long hours, the villagers were at their mercy.

All you could see were shadows, punctuated by harsh voices and blurred words. As the voices and footsteps drew closer and closer, my mother frantically signaled to us to hide beneath the bed and stay silent. My brother Bassam and I, both teenagers, desperately wanted to help our friends and neighbors. Bassam begged for a gun, but my mother had locked the door to the basement where the guns were stored and hidden the key. The attackers did not enter the house in which we had taken shelter, perhaps deterred by resistance; several neighbors stood their ground and were fighting back. A few hours later we found one of these men, Elias Hanna, lying in a pool of blood on the roof of the house facing ours. He was clutching a primitive hunting rifle.

Minutes of uncertainty seemed an eternity. All we could do was wait and pray. After gunfire had died down, we heard children and women crying and knew that the nightmare was over—for now. We ventured out cautiously. Everywhere fires were devouring houses and a thick black smoke blanketed the early morning sky. Destruction lay all around. Bassam and I raced down the hill to check on my father and grandparents, who had stayed in our house in the center of the village. They had survived. Several of our uncles, cousins, and friends were not so fortunate. Later we learned that the attackers had taken full control of the village, their thirst for revenge fueled further by the loss of a number of their comrades during the raid.

The police and the Lebanese army were nowhere to be seen. Not until seven o'clock in the morning—a full hour after the attackers had left the village—did an army unit appear on the outskirts of Tal Abbas. They had watched the killing and looting but had done nothing until it was clear the militiamen had gone. They tended to the wounded and collected the dead for burial. Their timing and weakness reflected the state of the country as a whole.

By mid-morning, all the dead had been brought to the village church for burial. The victims, old and young, lay on the blood-stained floor of the church, a simple but beautiful sanctuary that had been built by some of the dead. The church itself had also been a target. Pews were shattered, windows broken, icons ripped down, the crucifix desecrated. The elderly priest had to conduct the burial mass while grieving for his own dead son and his grandson, my best school friend. His voice choking, he blessed the victims while reciting their names. His calm demeanor and dignity gave us some peace. The mass was unusually short because the victims' wounds and conditions dictated immediate burial, and because some distraught inhabitants were in a hurry to leave the smoldering village.

Either by design or default, the Muslim extremists who had attacked our village were in league with Christian militias. Both camps opposed coexistence and diversity; both sought to intimidate their own coreligionists into believing they possessed a popular mandate. Even so, the majority of Muslims and Christians did not participate in the atrocities. What happened in my village is illustrative. The armed mob that killed, burned, and looted did not include most of our Muslim neighbors. In fact, just before the attack started, someone pounded on our door. Two Muslim messengers had been sent to warn my father what was coming. "It is out of our hands," they told him. "We tried and failed to stop them." They urged us to leave immediately and promised safe passage.

I remember so well the tone of their voices—sorrowful and fatalistic. The conversation was unbearably tense. The messengers stood near the door and refused the offer to come sit down. It was clear they feared for their lives as well as ours.

Our encounter with our Muslim neighbors that night was not unique. Other Christian villagers reported similar experiences. Some who had fled just ahead of the armed mob were rescued by Muslim neighbors and given refuge in their homes. After the smoke had cleared, our Muslim friends and colleagues came to express their shock and outrage. "We feel guilty and ashamed by what happened to you," an elderly friend told my grandfather softly. These

Muslim villagers reassured those inhabitants who left for the Christian areas in Beirut that their property and lands would be protected. Before the attack on the village, our neighbors had even offered to send their own young men to guard the village, to prevent outside criminals and militants from exploiting the breakdown in law and order.

The *muktar,* "mayor," and the village elders replied that there would be no need for such dramatic measures. Despite the pain, a sense of solidarity and shared loss had somehow prevailed.

⊂═◆═⊃

Christian and Muslim hard-liners alike intensified ethnic and religious cleansing throughout Lebanon, expelling entire communities from their homes and neighborhoods, but it was the Christian militias that pioneered the technique to consolidate their grip on their own regions. When Muslim extremists and their Palestinian allies retaliated, they played directly into the hands of the Christian militias, supplying them with a steady stream of foot soldiers and recruits.

Hicham, a militiaman with al-Jama'a al-Islamiya, was close to the center of the Muslim operations. "A friend of mine," he told me, "then a commander in the Muslim militia, thought of a genius reprisal against Christians: shelling the heavily populated Armenian quarter near Beirut, Borj Hammoud, where Christians least expected shelling." Assaad Chaftari, at the time a commander of Christian Phalangists, admitted that he would call movie theaters in West Beirut (predominantly Muslim), warning them that a bomb was planted there and causing a stampede. Assaad would then order his unit to shell the streets around the theaters, to kill as many Muslims as possible. "My friends and I wanted to mow down Muslims in any way. One thought we had was to reduce their fertility, especially since Muslims were outbreeding us. So we had planned to dump chemicals in the Muslim areas' water supply system to kill them en masse. Only technical problems stopped us." Assaad, a Greek Orthodox, recalled confessing to a Catholic priest that he had killed hundreds of Muslims. From the other side of the confessional screen, the priest responded: "My son, you are forgiven. Make it five hundred and then come back for confession." Assaad rose to the rank of lieutenant with the Lebanese Forces. Frustrated with the internal war among Christian factions in 1988, he gave up the killing business and started to work for interfaith reconciliation in Lebanon. He's now an electrical engineer and school supervisor.

But religion was merely a sideshow in this tragic drama. The Lebanese Civil War did not pit mainstream Muslim against mainstream Christian. On both sides those who led and sustained the fighting came from the fringes; they saw the war purely as an opportunity to seize power. Sectarianism, communitarianism, and *asabiya*, or group and tribal solidarity, were ways of grabbing a bigger share of the pie, meaning control of local and national bureaucracy. Tal Abbas may have been poor, but Lebanon was far from impoverished. Endowed with fertile land, beautiful beaches and mountains, the best universities and hospitals in the Arab world, it was a prize worth struggling for.

The war within the tribes was as brutal, if not more so, than the war between the tribes. Internecine warfare paralleled sectarian strife and sometimes even overshadowed it. "Comrades in arms turned their guns against one another," Hicham told me, whenever there was a lull in the fighting between Christians and Muslims. "Allied factions who had fought against common enemies—Christians—bickered over control of commercial quarters and spoils of war. Petty rivalries led to bloody fights over turf. My only brother, Toufic, twenty-three years old, Lebanon's champion in bodybuilding, was killed in battles among leftist and nationalist groups in West Beirut. My Muslim brothers and I had to make difficult choices. My endeavors to settle differences between two rival factions often failed. For example, I tried but failed to bridge the gap between my brother's faction, al-Murabitun, and the Nasserite Socialist Union—both of which are leftist and pan-Arab nationalist organizations. What was painful was that Toufic was with al-Murabitun, while my cousin and friend were with the Nasserite Union," Hicham said, tears in his eyes.

Hicham neglected to mention that he had belonged to an Islamist group that was at odds with both his brother's and cousin's factions. His story is typical of his generation, scarred by political violence and ideological mobilization. The Shiites were no less immune to internecine struggles. The two main Shiite militia—Amal, "Hope," and Hizbollah, "Party of God"—fought pitched battles across Beirut. "Sometimes skirmishes lasted for days and we were stuck at home with no food and provisions," Hicham said. In 1988 alone, fighting between Amal and Hizbollah resulted in the deaths of somewhere around three thousand people.

The Lebanese Civil War forced a shift in the balance of social forces within each community that was more dramatic than the one it forced between communities. The Shiite community, the largest in Lebanon, provides an example. Historically, a few feudal families, like the al-Ass'ads—who owned fertile agricultural land—dominated Shiite politics and business. By the end of the war, however, the feudal families had lost their privileged status and had been replaced by a new class of warlords and arrivistes. Power relations also shifted among the Sunnis. Muhieddine Itani, a former militiaman and friend of Hicham's, explained how the war elevated simple racketeers like Ibrahim Qoleilat into power brokers. "Before 1975 Ibrahim Qoleilat had been a tobacco smuggler," he said. "By the early 1980s, he had become one of the most influential Muslim leaders in West Beirut, leader of al-Murabitun, a pan-Arab nationalist militia. Cabinet ministers and businessmen knocked at his door to ask for favors. Under the pretext of defending his Sunni community, Qoleilat received weapons and assistance from the PLO and the Libyan government and protection money from businesses in West Beirut." Hicham also spoke of how some of his formerly impoverished Islamist cohorts secured a steady income from Iran, which poured money, arms, and ideologues into Lebanon. "They opened malls in Muslim areas in Beirut and its suburbs and made fortunes. They convinced Muslim consumers that their malls were 'Islamic' and that they cared about the well-being of the community. A nouveau riche class emerged out of the civil war."

My brother Bassam, a colonel in the Lebanese army, was killed in 1990 by the Lebanese Forces. The group had been trying to tighten its sectarian grip on a small strip of Christian territories in East Beirut, as well as the surrounding suburbs and mountains. They wanted to rid the area of the last remnants of state power—including Christian units of the Lebanese army. The militia fought against Christian opposition more fiercely than against putative Muslim enemies. Thousands of opponents were eliminated in order to cleanse the Christian heartland of critics and moderate voices. There was no space for dissent or disagreement.

Bassam's death left my family numb. We had thought by then that the worst was over. He was an exceptional young man. He could have joined me in the United States to do graduate work in physics or mathematics. Or he might have started painting—art was his passion; he left behind dozens of portraits. Instead he made a fateful decision to join the military as an officer, believing that the army was the only institution capable of bringing all of Lebanon's sects back together under one umbrella. He thought he could make a difference in the future of his country. "The army," I remember him saying, "could serve as the most effective force to unify the country and arrest political disintegration."

Bassam understood that Lebanon's divisions did not fall simply along Christian-Muslim lines but that they were based on more profound class and sociopolitical realities.

Throughout the civil war, Christian and Muslim militias specifically targeted rank-and-file soldiers. If the Lebanese central government recovered and rebuilt its institutions, the militias knew that they would be out of business. Anarchy sustained their power. From the onset, Bassam's predominately Christian unit, which was stationed in East Beirut and the mountains, faced persistent attacks by the Lebanese Forces and the Phalangists. The Lebanese army's future seemed grim: either it would be forced to serve the sectarian wishes of the warlords or it would be split up along the country's religious divide. The army resembled more a police apparatus than a professional fighting force. Since winning independence from France in 1943, Lebanon's politicians feared that a powerful army could endanger the country's fragile democracy and open society. "Lebanon's strength lies in its weakness," they declared, and believed.

The political establishment got it partially right: all around them in the 1950s and 1960s they watched Arab army officers stage coups d'état against constitutional governments and install themselves as absolute rulers. The most famous was Nasser, of course. On July 23, 1952, Colonel Nasser and a clutch of like-minded officers toppled Egypt's semi-constitutional monarchy and established a military dictatorship. By virtue of his charisma and oratory skills, Nasser popularized and institutionalized militarism in Arab politics. Egypt became a model. In quick succession Syria, Iraq, Sudan, Algeria, Libya, and others fell. Since those days, the man on horseback has dominated the Arab political scene and consolidated power in his hands. The preponderant role of the military and mukhabarat in Arab society and government is largely responsible for the lack of democracy and economic failure. Militarism and authoritarianism have grown into cancers.

While the Lebanese founding fathers might justifiably have feared a powerful army, they neglected the sine qua non of all nation building, namely an effective military. Civil society and social and sectarian harmony cannot survive without stability and security. To its citizens, the Lebanese army was simply a joke, better trained, it was said, at ballroom dancing than at fighting. When the civil war broke out it could neither restore peace nor subdue the militias.

As a nonsectarian officer in a fragile institution, Bassam knew the risks and expected the worst. He told me that he'd been approached by Christian warlords and was asked to join the Lebanese Forces in return for financial and political rewards. He was also warned that were his unit ever to challenge the militia's authority he could be killed. He lived in constant danger, but he was willing to do so for the sake of a diverse, multiethnic, multireligious society in the heart of the Arab world.

For my family, this was no consolation. We felt that Bassam's death had achieved very little; it had neither rekindled the flames of tolerance and coexistence nor had it brought the country closer to real democracy. Bassam's senior commander, along with dozens of junior officers, visited us to offer their condolences. When one officer tried to console my father, saying, "Bassam died so that Lebanon could live," the grieving, frustrated senior commander barked,

"Fuck Lebanon!" The sectarian warlords had split the spoils and anointed themselves as the country's new rulers. Bassam, like hundreds of thousands of individuals, was but another casualty. A small footnote in Lebanon's bloody history had left a big hole in our hearts.

After Bassam's death we learned that there had been earlier attempts on his life. Christian militias first used carrots and sticks, then blood and iron, to convert internal opponents to their cause. They were pioneers in the practice of putting religion in the service of political ambition. The Muslim jihadists learned well from their Christian counterparts. Though this is now largely forgotten, Christian fundamentalism is partially to blame for fueling Muslim militancy. Lebanon's Christians killed and pillaged in the name of the cross; their tactics were not lost on Hicham and Kamal's generation.

As the nation's frail legal institutions began to break apart, a primitive form of tribalism displaced civilization. Lebanon lost its moral compass. Religious coexistence gave way to estrangement and suspicion. Waving holy banners, neighbor railed against neighbor. People seized upon their communal identity in a desperate effort at self-preservation. The state of war pushed people into their sectarian bunkers and turned an open, tolerant society into a jungle.

Activism and militancy have afflicted all religions for centuries and centuries, long preceding the existence of what we call the Middle East, which is a twentieth-century colonialist invention. However, what happened beginning in the mid-1970s in Lebanon heralded a new era in Arab and Muslim politics. Politicized religion replaced secular nationalism as a dominant current in Muslim societies. Lebanon's breakdown bred a militant, fundamentalist fervor that moved with lightning speed. Christian fundamentalism, which was xenophobic and supremacist, fed into parallel tendencies in the Muslim camp. More than a hundred thousand people perished in the Lebanese Civil War. A million people—a third of the country's population—were displaced.

Moreover, as the Lebanese conflict grew it intersected with the 1979 Islamic revolution in Iran and the rise of the Ayatollah Khomeini, the architect of his country's transformation from a secular monarchy to a theocracy. But Khomeini's ideological program transcended Shiite Iran (of the 1.2 billion Muslims spread throughout fifty-three countries, roughly 10 percent are Shiites) and encompassed the ummah as a whole. Shortly after his triumphant return from fifteen years' exile in Iraq and a brief stopover in France, Khomeini called for dissolving the sectarian and national walls separating believers. It was time to reunite the Islamic community: "Boundaries should not be considered as the means of separation of the school of thought. Not only does Islam refuse to recognize any difference between Muslim countries, it is the champion of all oppressed people." Khomeini made his revolutionary ambitions clear: "We shall export our revolution to the whole world. Until the cry 'there is no God but God' resounds over the whole world, there will be struggle...Islam is the religion of militant individuals who are committed to truth and justice. It is the religion of those who desire freedom and independence. It is the school of those who struggle against imperialism."

Sunni and Shiite activists in other countries viewed Khomeini's revolution as a model. Kamal and his generation felt that if Khomeini, living in exile, could overthrow the most powerful dictatorship in the Middle East, they could do it, too: "Ayatollah Khomeini taught us a significant lesson—political will and charisma could overcome the *taghut* ['tyranny']." Hicham elaborated further: "By swiftly toppling the Shah of Iran, Khomeini boosted our morale—young Muslim activists—who dreamed of implementing shariah; he also mobilized moderate young Arab nationalists who were skeptical about the possibility of reestablishing the caliphate in the twentieth century." For both Shiite and Sunni Muslims, the Iranian revolution was nearly an apocalyptic event. Hicham, who is a Sunni, recalled, "My friends and I talked for days about the implications of the success of the Islamic revolution in Iran, and looked for ways to emulate [it]: what we should do to support it, or import it into Lebanon. Many of my friends, who were lay Muslim activists, after seeing 'the humiliation of the West' at the hands of Muslim mullahs in the Iranian revolution, they got rid of their western clothes, and put on Arab or Islamic attire."

In particular, the Islamic revolution in Iran galvanized the Shiites of Lebanon, who by then had become the country's largest constituency—almost 40 percent of the population. Khomeini's ideas and slogans spread with particular speed among the young and poor Shiites living in southern Lebanon and in the Bekaa valley, thanks to "ideological seminars"—accompanied by films—organized by the Iranians and their local allies.

Young Shiites also drifted increasingly to fundamentalist militancy because of Israel's constant incursions into their neighborhoods and homes. When Israel invaded Lebanon in June 1982 and later occupied Beirut, Khomeini dispatched one thousand Revolutionary Guards, or Pasdaran, to the Syrian-Lebanese border to confront the Israeli invader. As it turns out, these warriors did little or no fighting against the Israelis, who were based thirty-five miles away. Instead, they focused on religious and political indoctrination of the Shiites around the ancient city of Ba'albak in Lebanon's eastern Bekaa valley. One Revolutionary Guard told a British reporter in a rare interview, "Our only goal is to Islamize the place, as the Imam Khomeini says; we have to export the Islamic revolution to the world. So, like any other Muslims, we have come here with the aim of saving the deprived."

"There are two [main] cultures: the West and the East, the U.S. and the U.S.S.R; Islam is a third, a balance," a former guerilla named Hamza akl Hamieh told Robin Wright, an American reporter with the *London Times*. A militiaman in his early twenties when Khomeini seized power, the tall and muscular Hamza became a Shiite legend in the Middle East for hijacking six airplanes and getting away with it. Hamza's hijackings were designed to bring back Imam Musa al-Sadr, spiritual leader to Lebanon's one million Shiites. An Iranian-born friend of Khomeini, al-Sadr had founded Amal in the early 1970s. Along with two companions, Musa al-Sadr had disappeared in August 1978, after reportedly meeting with Libyan leader Colonel Mu'ammar Qaddafi. Al-Sadr had been on a tour of Arab countries, to appeal for aid and political support for the Shiites caught in the middle of the Israeli-Palestinian fight in south Lebanon. Libya was to have been his final stop.

"I see the future of my people must be determined by blood," Hamza said shortly after he was officially appointed military commander of Amal in 1984. "Our leaders in history said we must refuse the tyrants and the inequality. We must fight together to help all poor men in the world. And we must all fight to go back to Jerusalem." Indeed, Hamza and his generation of Shiites fought not only to gain the release of al-Sadr, but to establish an Islamic government in secular Lebanon. When pressed by an American reporter, Hamza said he favored Islamic rule throughout the Arab world, not just in Beirut, "because Islam makes equality to all the people, true Islam, not the kind in Saudi Arabia." Hamza and his collaborators were inspired by the "purity" of the Khomeini's revolution, as well as by his opposition to Western and communist influences. Khomeini's defiance of both the United States and the Soviet Union was intoxicating: "We must settle our accounts with great superpowers and show them that we can take on the whole world ideologically, despite all the painful problems that face us."

Khomeini's soldiers in Lebanon turned their guns against the Christian militias in the early 80s; they also launched an organized offensive to expel Israeli occupying troops, along with the Americans and Europeans who had come to separate the combatants and secure the peace. Shiite militants pioneered the practice of suicide bombings. On April 18, 1983, a powerful suicide bomb ripped through the U.S. embassy compound in Beirut, setting off a huge fire and killing sixty-three people. Five months later two car bomb explosions devastated the U.S. Marines command center at Beirut's international airport and a command post of the French contingent of the Multi-National Peacekeeping Force. At the time, it was the worst disaster for the American military since the Vietnam War: 241 marines and navy personnel were killed, dozens more left crippled for life. Fifty-eight French paratroopers also died.

Telephone calls by an unknown group called Islamic Jihad claimed responsibility: "We are the soldiers of God," said a voice in classical Arabic on behalf of the mysterious group. "We are neither Iranians, Syrians, nor Palestinians, but Muslims who follow the precepts of the Qur'an." Long before Osama bin Laden called for expelling the crusaders and Jews from Muslim lands, Islamic Jihad demanded that all Americans leave Lebanon or face death: "We said after that [embassy bombing] that we would strike more violently still. Now they understand with what they are dealing. Violence will remain our only way," the caller concluded. More suicide attacks against Israeli and American targets followed, along with the kidnapping of American and Western civilians. Islamic Jihad vowed "that not a single American or French will remain on this soil. We shall take no different course. And we shall not waver."

As Hamza's generation saw it, the United States supported their Christian and Zionist tormentors; now it had to pay for its crimes. "If America kills my people, then my people must kill Americans," announced Hussein Musawi, a pro-Iranian extremist who split with Amal and formed the

Islamic Amal party. "We have already said that if self-defense and if standing against American, Israeli, and French oppression constituted terrorism, then we are terrorists. This path is the path of blood, the path of martyrdom. For us death is easier than smoking a cigarette if it comes while fighting for the cause of God and while defending the oppressed." Musawi, sometimes called the Shiite "Carlos"—referring to Ilich Ramírez Sánchez, a notorious Venezuelan-born international terrorist also known as the Jackal—vehemently denied responsibility for the attacks on the Marines and French paratroopers, though he heaped praise on the martyrs who carried them out. In an interview with Robin Wright in the *London Times* he said, "I salute this good act, and I consider it a good deed and a legitimate right, and I bow to the spirits of the martyrs who carried out this operation."

Throughout the 1980s Khomeini's soldiers of God—led by Musawi's Islamic Amal, Islamic Jihad, and Hizbollah, a militia group established by militant Shiite clergy and modeled on a militia organization of the same name in Iran—turned large parts of Lebanon into jihadism's central headquarters. And like their Christian counterparts, Shiite jihadists in Lebanon promoted martyrdom as the most effective means of fighting local and international enemies. Al Qaeda, which is predominantly Sunni, was watching the unfolding struggle in Lebanon closely.

For example, when he first claimed direct responsibility for the attacks of September 11, Osama bin Laden linked them with the American-backed Israeli invasion of Lebanon. He recalled the horrific images of the siege of Beirut, the repeated attacks on West Beirut's high-rise apartment blocs, and the massacre of the innocent: "As I looked at those destroyed towers in Lebanon, it occurred to me to punish the oppressor in kind by destroying towers in America, so that it would have a taste of its own medicine and would be prevented from killing our women and children." After the bombing of the marine compound, rumors had it that American troops were preparing to overrun Beirut's southern suburbs, Hamza's neighborhood. He was defiant. "We are all in a race to see who goes to God first. I want to die before my friends. They want to die before me. We want to see our God. We welcome the bombs of Reagan," Hamza told CNN, smiling. Earlier he had explained, "They are depending on their good weapons. But they must know our people depend on good faith."

All this seemed to most outside observers—and certainly to most Americans—as religious fanaticism. The reality was far more complex. Since the 1970s religious fervor in the Arab world has fed into rage over economic, social, and political impotence. More and more Muslims turned to religion for spiritual sustenance and as a refuge from and a response to the seemingly unstoppable Westernization of their societies and oppressive political authoritarianism. Signs of this fervor could be observed almost everywhere. The number of men and women attending mosques jumped by huge amounts, even in previously secular countries such as Turkey and Morocco.

On Fridays in Cairo, Tehran, Karachi, Sana'a, Gaza, and Khartoum, you will find worshippers flooding the streets outside mosques. Today in formerly secular Arab cities, including Beirut, Damascus, and Algiers, most women and young girls are veiled (estimates place the number at 70 percent region-wide). Religious titles are estimated to account for almost 50 percent of all books published while the general publishing industry starves for readers. Visit any general bookstore in a Muslim country and you can't help but be impressed by the variety and quantity of religious texts. One Arab publisher told me that he couldn't compete with religious publications because of widespread public demand and the generous subsidies given to them by Saudi Arabia and other Gulf states.

Nonetheless, there is no direct correlation between religious fervor and militancy, despite what Western writers and journalists would like us to believe. Nor is there a direct correlation between religious fervor and political Islam, despite what Islamists and jihadists would like us to believe. A host of reasons underlie the rise of religious fervor, which has, after all, swept through the world, not merely Muslim lands.

As regards the Arab world, however, one thing is clear: pious middle-class Muslims have had little to do with militancy. They tend to be socially conservative and do not represent likely candidates for bin Laden's forces. One of the major reasons jihadists have failed to bring down the region's military regimes is that they have lacked the support of the Muslim middle class. Egypt and Algeria provide prime examples. During the 1990s both countries faced insurgencies that cost more than 150,000 lives. Militants waged all-out war against the authorities only to be decisively vanquished. Equally devastating, they lost the battle for the hearts and minds of mainstream Muslims. Citizens were repelled by the indiscriminate terrorist methods—the slaughter of policemen, intellectuals, and tourists. Of the hundreds of activists, students, and opinion makers I have spoken to few had anything positive to say about their oppressive regimes. But most also expressed unequivocal opposition to the ideology and tactics of jihadists, whom they view as reckless adventurers who could bring the temple down on everyone's head. There is little love lost between pious Muslims and the soldiers of God.

Thus, although manifestations of religious fervor continue to dominate Muslim societies, we should not conflate fervor with militancy. Muslim piety does not inculcate intolerance, nor incubate radicals and terrorists. The overwhelming majority of practicing Muslims are god-fearing citizens whose overriding goal is to take care of their families. Yet demonization of Muslims, which began in the 1970s, has reached new heights in the West. A sampling of recent book titles provides evidence enough: *The Age of Sacred Terror; Islam and Terrorism; The Blood of the Moon; Sword of Islam; Extreme Islam;* and *Religion of Peace or Refuge for Terror?* These books lump Islam, Muslims, Islamists, and jihadists together as a monolith, constituting a threat not merely to Western nations but to Western civilization itself. Muslims have become the New Barbarians.

I know this is not so because growing up in Tal Abbas offered me a different view. Though perched atop a hill in northern Lebanon, it existed in the vast common ground that lies between the great religions rather than at their extremes. So much was lost on that fateful October day when the village was attacked, but many of those who survived have, like Hicham and Kamal, embarked on the journey back to the center. The center did not hold, but neither has it disappeared.

Nonetheless, Tal Abbas never fully recovered. Young people who could afford to stay in the Christian sectors of Beirut did so. A decade after the attack, the inhabitants of the village were mostly elderly. Two decades later, a quarter of the population had returned. Today, three decades later, there is a fragile harmony, one not yet strong enough to withstand a new outbreak of political and social turmoil.

III

T𝔥𝔢 WARRIORS 𝔬𝔣 GOD:
THE SECOND GENERATION

"MY FULL NAME is Nasir Ahmad Nasir Abdallah al-Bahri, and I am an inhabitant of the Yemeni governorate of Shabwah. I am an expatriate of Saudi Arabia." So begins the startling *apologia pro vita sua* of a man who spent four years as Osama bin Laden's personal bodyguard—who crawled across sun-scorched rocks while at an Afghan guerilla training school; who, according to his account, struggled against the seduction of the blond seductresses of Bosnia; and who claimed personal responsibility in the attack on the USS *Cole* in Yemen on October 12, 2000, that killed seventeen American sailors and injured thirty-nine.

Al-Bahri, or, as he prefers to call himself, Abu-Jandal, is no sociopath, much as some Westerners might want to portray him as one. He's a man of uncommon intelligence, eager to learn about the world, and well-informed in his consideration of global politics. Early on in his life he became deeply distressed by the abuses he saw growing up with his Yemeni parents in Saudi Arabia. "My father was a mechanic in an engineering workshop, an ordinary man. My mother is a housewife." Like many Yemeni immigrants to Saudi Arabia, his family's prospects were limited. He received the same primary and secondary education as other Saudi boys, at the hands of the ultraorthodox Wahhabi religious scholars who have dominated Saudi schools for the last century. ("Wahhabi" refers to the sect inspired by the eighteenth-century Saudi Sheikh Mohammed ibn Abd al-Wahab, who wanted not simply to purify Sunni Islam but to wage jihad against all other forms of the religion.) The difference between Abu-Jandal and his schoolmates is that while they followed the conventional track—finding modest jobs, praying at the mosque, marrying, and having children—Abu-Jandal left home after he turned twenty-one in search of a grander, more meaningful existence, one that might align his life with the spiritual vision outlined by his teachers.

Abu-Jandal spent four years—from 1996 to 2000—by bin Laden's side, protecting him and performing sensitive missions for him. He was part of bin Laden's inner circle, was particularly close to his family, and was often present when operations were discussed and decisions made.

These days, Abu-Jandal spends his days in his small library in Sana'a, the capital of Yemen. In a photograph that shows him pulling a book down from his shelves, he might pass for an Oxford medievalist, even a young rabbi immersed in the finer points of the Torah. His neatly trimmed beard bears no resemblance to the warrior look of his old

Afghan comrades. I'm told that if you passed him on the street, you'd likely suppose he was a gentle, benevolent family man, his eyes sparkling with humor, the corners of his mouth turned upward in the manner of a good-natured uncle. But if you looked closely you would begin to perceive a peculiar tension between his eyes and his bearing—the way he stands and moves suggest something intimidatingly resolute. You can see this contradiction in the photographs taken of Abu-Jandal dressed in the traditional Yemeni garb: a turban, a white robe, a scabbard-sheathed dagger tucked inside a wide black belt, the kind of dress imprinted on the Western mind from Osama bin Laden's videotaped statements and threats.

Abu-Jandal represents what I would call the second generation of jihad. In a series of interviews conducted by the London-based Arabic-language newspaper *Al-Quds al-Arabi*, he retraced his steps after leaving home: the progressive loss of his innocence and his transformation from a spiritual activist into a hardened militant of the bin Laden network. The newspaper tends to be critical of American foreign policy, yet it took *Al-Quds al-Arabi* several months and intense lobbying to come to an agreement with the Yemeni authorities to talk with Abu-Jandal. The newspaper scored a big media coup by publishing his reminiscences.

I had no such luck. I contacted friends and colleagues in Yemen and asked if they could help find a way for me to meet and talk with Abu-Jandal. "Don't even try," they replied. Notorious for their secrecy and ruthlessness, the Yemeni security services kept the media away from Abu-Jandal. He knows too much about Al Qaeda and its contacts worldwide, including the Yemeni mukhabarat.

At first glance Abu-Jandal's journey mirrors that of Kamal. Both grew up in modest, devout households in backwater towns. Both were driven by a keen intelligence and were touched early on by spiritual passion. But while Kamal's generation was locally inspired and oriented—they wanted to restore their own nation to Allah—Abu-Jandal's dream defied national borders. He and his generation closely followed political developments in Muslim countries and were moved by the plight of persecuted believers worldwide. It was not some abstract idea of jihad that motivated Abu-Jandal but actual events affecting Muslims everywhere. While he never speaks openly of his emotional development—there is no "post-Freudian" tradition of internal self-reflection among Islamists and jihadists—what kept recurring to me about Abu-Jandal was what life for eighteen-year-old boys in the barren towns of Saudi Arabia must have been like. It could surely not be confused with the banal hijinks of Hollywood teens, yet like all adolescent males he was susceptible to all the sexual and apocalyptic fantasies that make male teenage life a battlefield.

To say all that may seem like little more than a truism, but we in the image-driven West all too often fail to see beyond the soft-focused newspaper images of turban, scabbard, and blade—just as all too many failed, a century ago, to remember that the fallen heroes of Flanders fields were pimply adolescent boys who would otherwise have been playing soccer in a back lot. In 1991 Abu-Jandal, like the

Catholic seminarian who has never kissed or held his sweet-heart's hand, was charged by powerful emotions with no ready target for their release—except for one thing: to par-take of the great awakening that had been launched by the new elders of political Islam like Kamal Habib. Unlike gen-erations of Arab boys before him, Abu-Jandal also enjoyed one other privilege we must never forget: The technology of mass communication had finally reached the Arab world. Is-lamic television matched with the proliferation of Islamic newspapers, made available by cheap offset printing tech-nology, broke down the walls of Muslim isolation. With the good news of Islamic triumph also came the fiery treatises and calls to battle from the Zawahiris and bin Ladens scat-tered across the Muslim landscape.

If European colonialism had fractured and emasculated the once-mighty Islamic Ottoman Empire, and if repressive nationalist regimes had suppressed free collective expression for the first three-quarters of the twentieth century, the civil wars in Algeria, Egypt, and Lebanon capped by the ebullient Islamic successes in Iran, Sudan, and Afghanistan, had cap-tured the attention of these newly literate young Muslim men as never before. Unlike their parents and grandparents, they were reading and watching television and listening to transis-tor radios. "We were interested in every matter that concerns Muslims, and we followed events affecting Muslims in Is-lamic newspapers and magazines," Abu-Jandal said. "We be-gan to follow events in Afghanistan, the battles in Khost, and, after that the fall of Kabul [to the mujahedeen]." It was as though wildfires of hope had broken out in every direction, from the Mediterranean coasts of the Maghreb through the valleys of the Fertile Crescent to the arid slopes of the Himalayas and even into the underbelly of Europe.

This new, educated, and hopeful generation felt much as the poet William Wordsworth and his fellow romantics did in the wake of the French Revolution, "Bliss was it in that dawn to be alive." But where, Abu-Jandal asked himself, to answer the call? It was neither imaginable nor possible to carry through in Saudi Arabia the campaigns his Muslim brothers were launching in, for example, Algeria. It was well understood that Saudi wealth had financed the ummah far and wide and that they owed their spiritual training to the Saudi government—the same regime that had also soiled their homeland by inviting the Americans to station their arms and soldiers on Saudi soil in the wake of the 1991 Gulf War. Jihad, however, could not be forgotten. And Abu-Jandal knew within his soul that he must follow its path. His choice was startling.

"My first station for jihad was Bosnia-Herzegovina." His reasoning was like that of all young radicals. "My journey for jihad...was an emotional trip," he said. Casting about for what he described as a mission of martyrdom, he ran away from home in October 1994. Lacking a passport, he entered Yemen on a visa and spent a year captivated by news reports of both the triumphs of and depredations on the ummah. The Muslim world, particularly the youth, were moved by the civil war in Yugoslavia that erupted in the early 1990s, following the death of Marshal Tito and the fall of communism. Via CNN they witnessed imagery of Bosnian Muslims becoming targets of a Serbian program of ethnic cleansing. Thousands of Muslim men, women, and children were killed; thousands of Muslim women were gang-raped by Serbian soldiers; and thousands of Bosnians were placed into concentration camps.

"I was watching the tragedies of Muslims in Bosnia; the slaughtering of children, women, and old people; the violation of honor and mass rape of girls; and the huge number of widows and orphans left by the war. Therefore, I decided to go to jihad as a young man who was raised on religious principles and chivalry and who is full of zeal about religion and care for Muslims."

Chivalry and zeal. Again, it is so easy for Westerners nursed on anti-Islamic diatribes by radio talk shows or over-blown treatises about war and civilization to turn these young men into two-dimensional caricatures. Change the names, the language, the dress, and we can see in Abu-Jandal the young volunteers of the Lincoln Brigade, heading off to defend democratic Spain in the 1930s, or from the opposite ideological scope "the contras" in the jungles of Nicaragua, lauded by President Ronald Reagan as "freedom fighters." However misguided we might think their specific choices, however painful it may be to see these men—who have attacked our own people here in our own cities—their motives were as universal as those of the Three Musketeers or the Boston Minutemen. Bosnia, a European land, presented a very special specter to this passionate Wahhabi soldier.

"The young people who had urged me to go to Bosnia were trying to make me change my mind and go instead to

Eritrea," he told his interviewer. "They said: We will facilitate your trip to any arena of jihad other than Bosnia because
there is moral corruption in Bosnia and many temptations
and European women." To Abu-Jandal's fellow jihadists, European Bosnia represented a sort of lush but lethal moral
jungle, vaguely like France might have seemed to the American doughboys of World War I. "The blond girls might
tempt you," they warned him. "I said: We do not want the
saying of the Prophet, God's peace and blessings be upon
him, to apply to us, when he answered a companion, who
had said: I fear for myself from the blond girls [as an excuse not to take part in jihad]." Going to Bosnia meant walking through the valley of the infidel, an act not unlike the
psalmist who would fear no evil so long as God was with
him. For this twenty-one-year-old's first venture into jihad,
and as he said, toward martyrdom, we can only assume that
the chance of seduction raised the ante of moral struggle
even higher. Of course to speak of such things is, for a Wah-
habist, all but inconceivable. But the implications are clear,
when a little further in the interview he reiterates the Wahhabi view of all women, which is that they are temptresses
who must be kept under strict segregation.

Nevertheless, Abu-Jandal was already sufficiently media-
savvy to understand that Bosnia, more than Afghanistan or
Algeria or Eritrea, had stolen the global spotlight, and therefore that it was important for the movement to carry jihad
forward there.

He discovered that Bosnia's Muslims bore about as much
relation to Wahhabi Islam as a Church of England gardener

JOURNEY of the JIHADIST

did to a Bible Belt evangelist. Mosques were scarce in Bosnia, and active five-times-a-day practitioners even scarcer. Abu-Jandal found himself assigned to guard duty in the Arab Mujahedeen Brigade, an all-Arab unit within the Bosnian army, consisting of some 450 fighters from Egypt, Algeria, Saudi Arabia, and Yemen. He apparently saw little if any combat. The Mujahedeen Brigade spent much of its energy bringing spiritual salvation to the souls of fallen Muslims. "Communist ideology had wiped out all the features of the Islamic religion and the understanding of Islam. We saw some Muslim youths wearing a cross around their necks without knowing what this meant, although they belonged to Muslim families and some of them had Arab and Muslim names. They were completely ignorant of Islam. Therefore, we saw that the responsibility we shouldered in Bosnia was broader and more comprehensive than the mission of combat, for which we had come. So we found that we became bearers of weapons *and* at the same time bearers of a call, a book, and a message."

In 1987 the first intifada, or uprising, against Israel's occupation of the Palestinian territories of the West Bank and the Gaza Strip erupted; Palestinians threw stones at Israeli troops, who responded with a disproportionate use of force. Hundreds of Palestinians, including children and teenagers, were killed, scenes of which were broadcast via CNN and other media worldwide. This was before the age of suicide bombers or the new Arab media—Al Jazeera. The first Palestinian intifada, which was non-violent, earned the Palestinians worldwide empathy; it also fueled the passions of young

Arabs and Muslims. Ironically, the first intifada gave birth to the militant Islamist movement Hamas in 1990. Coupled with the 1991 Gulf War, these internal developments contributed to the Israeli-Palestinian peace negotiations in Madrid and Oslo in the early 1990s.

There was of course more to Abu-Jandal and his generation's story than their emotional reaction to the persecution of Muslims in Bosnia or the intifada in Palestine. In the 1980s there already existed a religious and ideological context that was hospitable to jihadist causes throughout Muslim countries. Radical clerics everywhere were encouraging young men to leave their families and homes to wage jihad in distant lands. By the close of the decade the campaign by the mujahedeen—another term for, and linguistically connected to jihadists, though its common usage suggests guerilla tactics—against the Russians in Afghanistan had transformed many volunteers into hardened warriors. What began as a defense against foreign invaders had become a global jihad offensive. It is essential to remember the centrality of the Russian and American roles in that transformation—the Russians as provocateurs and the Americans as co-bankers, with the Saudis, of the Afghan volunteers.

Kamal remembered that in the 1980s thousands of Egyptian Islamists went to Afghanistan to gain combat experience and learn insurgency techniques in preparation for the fight at home. In his 2001 diaries, Zawahiri confirms this: "The Afghan arena, especially after the Russians had withdrawn, became a staging theater of jihad against renegade [Muslim] rulers who allied themselves with the foreign authorities."

Abu-Jandal confessed that his motivation to wage jihad drew on more than chivalry and courage. Having undergone his spiritual awakening at age fourteen, he was especially susceptible to the religious campaigns articulated in the Friday sermons, the tape cassettes, the magazines, and other media. All the while he was reading and listening. "I wished I was one of those mujahedeen, defending Muslim lands." When he was ready to act, the Saudi ulemas were ready to take advantage. Materially and morally, they were preparing young Muslims to migrate and do battle on behalf of the ummah.

Abu-Jandal cited a hard-line cleric, Salman al-Awdah, one of bin Laden's spiritual gurus, who was highly active in recruiting and equipping many youths to go to Afghanistan, Bosnia-Herzegovina, Tajikistan, and elsewhere. While in the 1980s al-Awdah used the pulpit to incite young Muslims against Soviet atheism, in the 1990s he turned his rhetorical guns against "infidel" Western nations, particularly the United States. In an article entitled the "End of History," al-Awdah didn't mince words: "The oppressors are the swords of Allah on earth. First Allah takes his revenge by them, and then against them. The same as Allah has used, in Islamist eyes, the United States in order to destroy the Soviet Union, so he will take revenge against the Americans by destroying them." In November 2004, Awdah, along with twenty-five prominent Saudi religious scholars, posted an open letter on the Internet urging Iraqis to support jihad against "the major crime of America's occupation of Iraq."

Actual funding of the mujahedeen, according to Abu-Jandal, came from private donations, with the religious sheikhs forming a critical link between the charitable donors and the young warriors. "For instance," he said, "I was equipped for my first jihad by a woman. She worked as a schoolteacher. She had heard about the tragedies that had befallen the Muslims in Bosnia-Herzegovina and wanted to contribute to their defense. She asked, 'What is the best thing I can contribute?' The answer was: equip a mujahid. She said, 'I will donate a full month's salary to equip a mujahid.' It was equivalent to approximately $2,000." (Interestingly, many Muslim women did the same thing during the Crusades in the eleventh through thirteenth centuries.)

Such donations, Abu-Jandal said, filled the coffers of mosques and sympathetic charitable foundations. "There were astronomical sums available for equipping the youths for jihad. All of Saudi Arabia—the government, the religious scholars, and ordinary people—was on the side of driving the youths toward jihad." Until the early 1990s, Abu-Jandal recalled nostalgically, the government and people of Saudi Arabia were in total harmony on jihad: "The Islamic climate was everywhere in Saudi Arabia, and the Islamic spirit was in everything: in the councils of scholars and in religious gatherings. I remember that many people used to come from outside Saudi Arabia, from Arab and Muslim states, in order to live in this Islamic environment. The entire society there was one fabric. It was impossible to find a house without the fragrance of Islamic trends, in any form. Thus, if a household did not have a young man who observed the faith, it had a young woman who observed the faith. If it did not have a young woman who observed the faith, the household perhaps had an Islamic tape or an Islamic book. Up until almost 1991 [when the first American troops were stationed in the kingdom], all such jihadist actions existed in the name of Islam and leaned toward Islam."

Everything changed after the 1990 Iraqi invasion and subsequent annexation of Kuwait. Like Saudi Arabia, Kuwait had backed Iraq during its war with Iran from 1980 to 1988; they never expected Saddam's regime to turn against them. Despite having rich petrodollar-based economies, these states had never invested in national defense in any meaningful way—they relied on the United States to be the guarantor of their national security. The 1991 Persian Gulf War exposed not only the vulnerability of the Saudi state but threw into question the regime's Islamic credentials. Educated Saudis from all walks of life began publicly questioning the government's policies. There was great commotion within religious circles over the notion of allowing a large number of non-Muslim troops in the kingdom. Osama bin Laden, who had returned from Afghanistan as a hero a year earlier, tried to use his influence within the highest echelons of the royal family to allow his legion of Arab and Muslim fighters to expel Iraqi forces from Kuwait and to defend the Saudi kingdom against possible Baathist aggression. The ruling family worried that this would endanger their survival; relying on "infidels" to survive led to their delegitimization in the eyes of many Saudis.

As the guardian of the two Holy sanctuaries—Mecca and Medina—Saudi Arabia was naturally positioned to be the beating heart of jihadism. Millions of Muslims make pilgrimages to these two sites each year. Moreover, as the biggest oil producer (nine million barrels a day) with the largest proven reserves in the world (more than three hundred billion barrels), the Saudi leaders have the resources to sell to Muslims everywhere. The revenue has allowed the Saudis to keep the religious establishment happy. The clerics, in turn, have used their resources to spread a puritanical, ultraconservative interpretation of Wahhabism/Salafism (Wahhabis prefer to be referred to as "Salifis," referring to the Prophet and his companions) across the globe. The Saudi religious establishment funds mosques, Islamic centers, and schools, supplies teachers, preachers, and scholars, and assists in the proliferation of its print literature. Wahhabist/Salafist theology views any form of intellectualism, mysticism, or diversity within Islam as corruptive and un-Islamic; it insists that on all issues Muslims refer to the original textual sources of the Qur'an and the Sunnah, as laid down by the Prophet and his companions and early descendents, the *al-salaf al-salih,* or "righteous predecesors."

At the core of Wahhabi/Salafi belief is an implacable opposition to women's rights. Wahhabis and Salafis enforce total segregation of the sexes. They would fight to death to exclude women from the public arena. As we've seen, female sexuality is viewed as dangerous and subversive. For many, the prospect of the liberated female poses an even greater threat than invasion by the infidel foreigner. Men, as Abu-Jandal was warned before shipping out to Bosnia, are simply too weak and too vulnerable to withstand Eve's seductive charms.

Nowhere was Wahhabi/Salafi orthodoxy sharper than under Taliban regime Afghanistan and especially within bin Laden's Al Qaeda movement. To understand bin Laden, Abu-Jandal explained, one has to know that he is a Salafi. "He was brought up with the Saudi Salafi educational line that rests on one approach—the Wahhabi approach. All Saudi schools teach the call of Sheikh Mohammed ibn Abd al-Wahhab. We used to be taught that call every morning and evening, in the same way mothers suckle their babies, until we came to know the personality of Mohammed ibn Abd al-Wahhab more than we knew about the history of the Prophet's companions. Sheikh Osama bin Laden emerged from under the cloak of the Saudi Salafi call." When he eventually became bin Laden's bodyguard, Abu-Jandal gained an intimate view into his boss's private domain: "His [four] wives lived in one house that had only one floor. They lived in perfect harmony. Sheikh Osama was firm in managing his household's affairs just as he was firm in handling matters outside his home. We never heard about any conflict among the wives." The virtual invisibility of bin Laden's wives, indeed the strict prohibition of anyone outside the family from even uttering their names in public, was characteristic of the utopian Islam the Taliban worked to create in Afghanistan, and hoped eventually to replicate throughout the Muslim world.

In the 1990s, bin Laden's Wahhabi/Salafi cohorts dominated the Saudi landscape, Abu-Jandal boasted. Soon they felt able to dispatch thousands of mujahedeen on missions around the world: Chechnya, Kashmir, the Philippines,

Somalia, Indonesia, Morocco, Egypt, Algeria, France, Spain, and England: "He [bin Laden] used to say: 'I have more than 40,000 mujahedeen in the land of the two holy mosques alone.'" Of all nationalities and ethnicities, "the sons of the two holy mosques" formed by far the biggest and most dedicated contingent within Al Qaeda; they spearheaded bin Laden's striking force. Fifteen of the nineteen hijackers involved in the September 11 attacks were Saudis chosen by bin Laden for their *asabiya*, "tribal loyalty." Tribal society is disproportionately more religious than other social configurations. Among members of a tribe, the good of the individual is subservient to the collective good. When overlapping with religious and ideological affinities, tribal loyalties can be a powerful motivator.

The Saudis' promotion of jihad played directly on Abu-Jandal. Not only did they finance his religious mentors, who in effect set themselves up as jihad recruiters, but the government-financed and -controlled media went out of its way to promote the glories of jihad. It organized press interviews with some of the leading mujahedeen figures, notably Sheikh Mahmoud Bahathig (Abu Abd al-Aziz), the first emir of the Arab mujahedeen in Bosnia, who gave a six-part interview in a Saudi newspaper. Abu-Jandal said he attended Sheikh Mahmoud Bahathig's lectures about the progress of the Bosnia jihad. Not long afterward, Abu-Jandal, supported by the schoolteacher's donation, decamped for Bosnia and joined Sheikh Mahmoud's all-Arab brigade.

The Saudi link to Taliban Afghanistan has been widely documented despite early denials of any connection to bin

Laden's cohorts. The Arabic-language *Al Hayat* serialized the memoirs of an Afghan jihadist named Abdullah Anas shortly after September 11. An Algerian son-in-law of Sheikh Abdullah Azzam, bin Laden's spiritual father, Anas wrote that the Saudis had donated millions of dollars to Azzam's *Maktab al-Khadamat*, "Services Bureau," which housed and trained thousands of Muslim volunteers in Peshawar, Pakistan, a center of protection and support for bin Laden and the Al Qaeda guerillas after the American bombing had begun. According to Anas, Saudi Arabia also became a ferrying port for Arab veterans and jihadists.

The Saudi's support for jihad expeditions abroad surely reflected their long-held Wahhabi convictions. (While the Saudi regime, for that matter the entire country, is Wahhabi, we should keep in mind that while all jihadists are to varying degrees Wahhabi in religious orientation, not all Wahhabis are jihadists. In fact, jihadists form a small subset of the larger Wahhabi community.) But the Saudi elites have never been naive in their assessments. A short boat ride across the Red Sea from Saudi Arabia lies Egypt, where the civil war mounted by Kamal and his allies had generated massive chaos among the newly educated and frustrated youth. The Saudis reasoned that if they could take credit for and harness that radical, idealist energy, then export it to a barren and faraway land, so much the better for homeland security. Saudi Arabian Airlines, the country's official carrier, even created a sort of jihad youth fare, offering a 75 percent discount on ticket purchases by any volunteer flying to Afghanistan to wage jihad.

Prodded by the United States, Saudi Arabia and Pakistan (which, despite its egregious human rights record, has long been America's primary client state in central and south Asia) led the way in supporting the jihad caravan to the Afghanistan war against the Soviets. Other Muslim states, among them Egypt, Algeria, Indonesia, Turkey, Morocco, and Jordan, contributed their shares after pressure from the U.S. According to a first-hand account by Sabri Faraj Ayman, an official from the Muslim Brotherhood who fought in Afghanistan, President Sadat himself met with the Brothers' leaders and encouraged them to help the Afghanis by sending volunteers. Faraj tells us that the Brotherhood happily accepted Sadat's public-relations gesture and continued sending volunteers even after his assassination in 1981.

⊱══✦══⊰

Afghanistan in the 1980s provides an Islamic parallel to the enormous tent and tabernacle gatherings of the Great Awakening in eighteenth-century New England, where ecstatic Christians gathered to proclaim and reassert their holy mission to build a "new world" guided by divine providence. Young Muslim men—and only men—poured into that sad, desperate, depleted land whose only reliable cash income came from opium poppies. But what started as a religious gathering quickly turned into a battlefield, one on which these passionate idealists—far from home and family— came to know and rely upon each other for their lives. Very soon they began to bury their new comrades, cut down first by Russian Kalashnikovs and later by American shrapnel.

With nothing to sustain them but the promise that, once martyred, they would be rewarded by God, they gave their lives to the defense of the ummah. Those who didn't lose their lives lost their innocence.

In Afghanistan was assembled the first truly global army of Islamic warriors—the Afghan Arabs. Never before in modern times had so many Muslims from so many different lands speaking so many tongues journeyed to a Muslim country to fight against a common enemy—Egyptians, Saudis, Yemenis, Palestinians, Algerians, Sudanese, Iraqi Kurds, Kuwaitis, Turks, Jordanians, Syrians, Libyans, Tunisians, Moroccans, Lebanese, Pakistanis, Indians, Indonesians, Malaysians, and others. For a fleeting moment in Afghanistan, in the eyes of Islamists and Muslims alike, there existed a community of believers united in arms against infidel encroachment and aggression. As a Yemeni veteran confided to me, "Afghanistan reminded Muslims of all colors and races that what unites us is much more important than the superficial differences wrought by colonialism, secular nationalism, and other material ideologies. We felt we were on the verge of reenacting and reliving the Golden Age of our blessed ancestors."

Afghan veterans admitted how much their war experiences had transformed them. As Sabri Faraj Ayman wrote in his diaries, "Initially I enjoyed violence but the longer I fought, the less pleasure I took in it, and then it became more of a psychological burden. At this latter stage I lost interest in life and desired death." Faraj recalled veterans competing against one another to see who would be martyred first and enter the heavenly kingdom of eternal peace. "Fighters had sweet dreams of fulfilling their duty to God and Prophet. Who could resist the magic of jihad and martyrdom and courage and sacrifice?...Who could resist the dreams of reestablishing the caliphate ...an Islamic state encompassing Muslims from Senegal to the Philippines?"

A warrior class unlike any Kamal's generation could ever have imagined, these mujahedeen were mobile and highly trained, willing and ready to do battle anywhere, anyplace, anytime. Their focus wasn't local but international. Among the thousands of young men inspired by them was the teenaged Abu-Jandal. "I remember that one of our colleagues went to Afghanistan and spent two weeks during the month of Ramadan. When he came back, we gave him a hero's welcome. When we used to look at the Afghan costumes worn by the returning mujahedeen as they walked the streets of Jedda, Mecca, or Medina, we used to feel we were living with the generation of the triumphant companions of the Prophet, and hence we looked up to them as an example and an authority."

While Abu-Jandal celebrated the mujahedeen, older veterans like Kamal were highly critical. "What you need to understand," Kamal told me not long after I had first met him, "is that the war in Afghanistan internationalized and militarized the jihadist movement further. Al-shabab, who fought the Soviet occupiers in Afghanistan, put religion at the service of war rather than the other way around. Afghan veterans paid little attention to society and balance of power. Violence replaced politics as a means of interaction. Afghan fighters committed a cardinal sin by overlooking a vital lesson: Politics is the continuation of war by other means." He shook his head in dismay. "Why did they fight if they could have achieved similar ends though diplomacy and negotiation? That would have been less costly and more effective. I am not talking about strategy but common sense."

In 1989 the Russians retreated from Afghanistan. Rather than disband and go home, thousands of Afghan veterans felt so empowered by having defeated one of the world's superpowers that they embarked on new militant adventures. Fighters and campaigners became unpaid mercenaries. It is a story as old as human history. This was certainly the case with the crusaders, some of whom turned to piracy even before they had reached the Holy Land. After the peace was won in Northern Ireland, rump elements of the Irish Republican provisionals turned into anti-Asian attack squads. Some even became shakedown operators or murder-for-hire hit squads. Revolutionaries who had devoted their lives to toppling the murderous dictatorship of Zaire's Mobutu Sese Seko themselves became dictatorial brigands. All too often those who become infected with the virus of violent mission find themselves unable to settle for the mundane. While their zeal tended to keep the mujahedeen from being corrupted, it also fed their taste for blood and victory.

In his memoirs, serialized in the Arabic-language newspaper *Asharq Al-Awsat*, Zawahiri, one of the first senior Egyptian jihadists to go to Afghanistan, boasts that "The jihad battles in Afghanistan destroyed the myth of a [superpower] in the minds of young Muslim mujahedeen. The Soviet

Union, a superpower with the largest land army in the world, was destroyed, and the remnants of its troops fled Afghanistan before the eyes of the Muslim youths and with their participation."

Even had they wanted to, many veterans could not return to their native countries, whose governments viewed them as security risks. Arab rulers were determined to keep these potential troublemakers as far from home as possible—to disown them or, better still, to make them somebody else's headache. The governments conveniently forgot that they had been encouraged and funded by America to join in the anticommunist Afghan campaign. That was in the early 1980s, when the Reagan administration had subordinated its foreign policy to defeating "the Evil Empire." The problem was that the genie of jihad would not go back in the bottle.

To Kamal's thinking, the Afghan experience infused the jihadists with hubris. Having forcibly evicted a superpower from Afghanistan, they could also defeat the corrupt, unIslamic regimes in their home countries. "Islamists overestimated their real strength and felt overconfident. Power went to their heads. They miscalculated." This was the period during which Egyptian militants waged a bloody campaign at home and were decimated. "This move set us back more than a decade."

This new warrior generation would not be welcomed home as heroes, but it didn't much care about coming home. They saw themselves, as Abu-Jandal says repeatedly in his interviews, as the vanguard of the ummah, not as citizens of discredited states invented by Europeans from the ashes of

Ottoman glory. "When we went forward [abroad] for jihad, we experienced reality. We saw things that were more awful than anything we had expected or had heard or seen in the media. It was as though we were like 'a cat with closed eyes' that opened its eyes. We began to have real contact with the other trends, the enemies of the ummah, and the ideology of the ummah began to evolve in our minds. We realized we were a state that had a distinguished place among nations."

Abu-Jandal admitted that when he left the Saudi peninsula for Bosnia as a young man he possessed no very profound understanding of the theology of jihad, nor had his generation undergone the sort of ideological and strategic education that university-trained Egyptian militants like Kamal had. He was no more than a highly motivated novice whose exposure to jihadist ideology and experience in military skills were at best superficial. Before having fought in Bosnia, he said, "I used to consider jihad and carrying arms a kind of voluntary work. I did not view jihad as a religious duty prescribed to every individual [*fard'ayn*, 'personal duty'] [as jihadists do], but as a *fard kifaya* ['collective duty'], i.e., if it is carried out by some, then others are exempt from it, albeit with their parents' consent." In Islam, there are two types of obligations: those that are incumbent upon each Muslim to fulfill individually, and those, which, if disposed by a few from the community/region/nation, absolve all others from having to do it.

By the time he departed from Bosnia a few months later, he viewed jihad as a permanent and personal duty, a pillar of Islam: "There is a dilemma and a misunderstanding that most Muslims face. It can be summed up in that jihad has become merely a matter of thought for them, and they forgot that jihad is something that God has prescribed to us as a religious duty, like prayers, fasting, alms-giving, and pilgrimmages."

The impact Abu-Jandal's sojourn in Bosnia had on him was far greater than the impact he had on events in Bosnia. He and his brethren learned the importance of organization. They realized that "no practical jihadist experiment can succeed except through organized action." It became clear to them that fighting against non-Muslims who threatened Muslim lands was not going to be enough. Volunteers could not simply behave like bands of fighters, roving from one area of the world to another. What was needed was a global network that could organize the volunteers and dispatch them to various parts of the world. As they were being scattered they took that insight with them. One group of fighters headed for the Philippines while another went to Chechnya. Abu-Jandal headed for Somalia. He decided on Somalia with no prompting from Al Qaeda or any other outside organization. Defending "the weak and oppressed in this world, and the love of martyrdom for that cause," he maintained, were his only considerations. "This was the only thing on my mind. I had not met Osama bin Laden face-to-face and did not know him until after I returned from Tajikistan to Afghanistan [in 1996]."

The majority of Somalis are Muslim, and jihadists close to bin Laden were able to set up shop in the country. In 1991, Somalian warlords ousted Muhammad Siad Barre, who had been ruling the country since 1969. The collapse of Barre's regime led to anarchy and violence and precipitated a humanitarian crisis. In December of 1992, the United Nations authorized a U.S.-led intervention—Operation Restore Hope—which culminated in August 1993, in a seventeen-hour battle—known to Americans as the "Black Hawk Down" incident and to Arabs as Ma-alinti Rangers, "day of the Rangers"—in which eighteen U.S. and one Malaysian soldier were killed and seventy-three wounded. Bodies of dead American soldiers were dragged through the streets of Mogadishu. Between 500 and 1000 Somali militia and civilians also died, although the official number has never been released. The entire battle was photographed and video-taped by satellite. On October 7, 1993, President Clinton responded by announcing the withdrawal of U.S. troops from Somalia. In an interview with CNN's Peter Arnett in 1996, bin Laden boasted how jihadist fighters fought the Americans in Mogadishu: "Those who fought in Somalia told us that they were surprised by how low the morale of the American soldiers was in comparison with the Russian soldiers they had fought. The Americans ran away from those mujahedeen while the Russians stayed. If the United States still brags about this kind of power, even after the successive defeats in Vietnam, Beirut, Aden, and Somalia, then let its troops go back to those who are awaiting their return."

"We did not go because of the Americans, but because there was a conflict between Muslims and others," insisted Abu-Jandal. However, he and his fellow travelers quickly discovered that they were less than welcome. Indeed they were forced to flee the country within a few weeks. "The Somalis treated us in a bad way and laid a media siege against us...I sensed a number of attempts by some leaders of the Somali Islamic movement to entrap us, although we had come to support them. We tried to convince them that we were messengers for the people behind us, but they did not believe us."

The Somalis turned out to be hostile to the foreign fighters, particularly the Arab volunteers. Abu-Jandal maintained that he never participated in any military operations against U.S. forces while in Somalia. "By the time we had entered Somalia, Al Qaeda organization itself had already started to leave the country." After the Americans had departed Somalia, there was nothing left for bin Laden's men to do, although a few of his senior aides, such as Abu Ubaidah al-Banshiri, stayed behind and established operations in the benighted country.

Inspired though ineffective in Bosnia, defeated in Somalia, Abu-Jandal needed to figure out his next step. Had he been ready to settle down, as some of his comrades were, he would never have crossed paths with Osama bin Laden. But the jihadist mission still burned within him. He joined a group of young men, most of them from the Arabian Peninsula, that was heading first to Tajikistan, then to Pakistan, and eventually to Afghanistan. This so-called Group of the North, consisting of thirty-six men, spent six fruitless months in Tajikistan, where they were apparently turned upon by forces allied with the Afghan leader Ahmad Shah Masud. "They betrayed us while we were on the way and tried to commit treachery against us several times." The former So-viet republic had become an independent state and was almost immediately torn apart by a civil war between the Tajik government allied with regionally based ruling elites, on the one hand, and disenfranchised regions, democratic liberal reformists, and a loose organization of Islamists known as the United Tajik Opposition (UTO), on the other. A UN-sponsored armistice finally ended the war in 1997. During the same period, Afghanistan was also in chaos, as the Islamist rebel alliance that had forced Soviet troops out of the country was engaged in an internal power struggle. One of the key groups was Jamaat-i-Islami, an organization composed predominantly of ethnic Tajik Afghans. They con-trolled the northeastern parts of Afghanistan, which borders Tajikistan. Most of the Arabs were aligned with Hizbi-i-Islami, the Pashtun rivals to Jamiat-i-Islami; it also had extensive contacts in Tajikstan and worked to thwart the Group of the North's efforts. Eventually the entire group moved on to the training camps in eastern Afghanistan.

⚔

In 1996, Abu-Jandal met Osama bin Laden, who was living in Jahalabad at the time. Bin Laden had first come to Af-ghanistan in the early 1980s as a guest of Sheikh Abdullah Azzam, the spiritual godfather to the Arab mujahedeen. Azzam set up the first guesthouse—through the aforemen-tioned "Services Bureau"—in Peshawar, Pakistan, to facilitate and coordinate the flow of volunteers to and from Afghani-stan. Bin Laden financed the Bureau's jihad activities while

Azzam managed its operations. In 1996, the Clinton administration and Saudi Arabia had pressured the Islamist regime in the Sudan, where bin Laden was living, to expel him. He had had nowhere else to go except Afghanistan, where he had maintained contacts from his days with Azzam. This was the year he made up his mind to fight the Americans, the year he issued his infamous fatwa, declaring "War Against the Americans Occupying the Land of the Two Holy Mosques"—meaning Saudi Arabia. Bin Laden began to aggressively recruit young men to his cause, particularly from his native Saudi Arabia and the Gulf states, like Abu-Jandal.

But bin Laden's first visit had been critical. More than any other contemporary figure, Abdullah Azzam, a Palestinian refugee born in the West Bank, exercised a formative influence over bin Laden. In Saudi Arabia, he had sat at Azzam's feet like an adoring college student, and indeed Azzam—who had done a Ph.D. in Usul al-Fiqh, "Principles of Islamic Jurisprudence," in 1973 at Cairo University—had taught Islamic studies at King Abdul Aziz University in Jeddah for many years. He outgrew his academic status and became a key player in mobilizing support for the Islamist rebels fighting the communist regime in Kabul. Azzam built a network that was scholarly, ideological, and paramilitary, and there can be no underestimating how much that network contributed to the globalization of Islamist militancy. After the Soviet invasion of Afghanistan in 1979, Azzam moved to Pakistan to teach at International Islamic University in Islamabad. Soon after he established the Services

Bureau, which was now running not merely guesthouses in Peshawar but also paramilitary training camps in Afghanistan to prepare international volunteers to fight alongside the Afghani rebels. During the 1980s, Azzam traveled throughout the Middle East, Europe, and North America—including some fifty U.S. cities—to raise money and preach jihad.

Azzam believed that the fight in Afghanistan would be the launchpad for a greater struggle whose ultimate objective was establishing a caliphate across all Muslim lands. Azzam's radical ideology, combined with his efforts to impart paramilitary training, led to the creation of a multinational legion of highly motivated and highly trained militants intent on spreading jihad throughout the Muslim world. "Jihad and the rifle alone: no negotiations, no conferences and no dialogues," was among his most oft-quoted sentiments.

By the end of the 1980s, as seasoned fighters from the Egyptian Tanzim al-Jihad and the Islamic Group joined the Afghan war against the Russian occupation, bin Laden developed new ties with their militant leaders, and his appetite and imagination expanded. They established an advanced training camp called al-Faruq, designed as a sort of military college to prepare senior officers to lead operations wherever they were deemed necessary. "If the jihad in Afghanistan were to end," as Abu-Jandal put it, "graduates of the college could go anywhere in the world and capably command battles. Those objectives were actually achieved through the success accomplished by the young men who had moved to many fronts outside Afghanistan, in Bosnia-Herzegovina,

Chechnya, the Philippines, Eritrea, Somalia, Burma, as well of course as the cells in Western Europe that brought down the World Trade Center on September 11, 2001."

The victories achieved on those fronts, Abu-Jandal boasted, were a direct result of the rigorous curricula offered at al-Faruqe, which in the 1990s became what we know today as Al Qaeda. Throughout the same period new volunteers were emerging, swearing allegiance to bin Laden and the "emirs" of two Egyptian factions. One was Zawahiri and his Tanzim al-Jihad. By 1990 Zawahiri had cemented his relationship with bin Laden.

By this point, however, bin Laden and Azzam had gone their separate ways. Unlike Zawahiri, who went to Afghanistan to prepare for a final showdown with the Egyptian regime and to call for the overthrow of pro-Western Muslim governments, Azzam opposed expanding the Afghan jihad against other Muslims. He was more concerned about transforming the Arab mujahedeen into an Islamic rapid response force to assist persecuted Muslims worldwide. His thoughts returned to his homeland; Afghanistan was merely a dress rehersal for jihad in Palestine.

In addition, Azzam's conception of jihad was more limited and defensive than Zawahiri's, even than Kamal's. In his sermons and writings, Azzam stressed that jihad was a personal duty (fard'ayn). "When the enemy enters the land of the Muslims, jihad becomes individually obligatory." Unlike Sayyid Qutb, Zawahiri, and other revolutionaries, Azzam emphasized resistance, not aggression. For example, in Afghanistan he had focused almost exclusively on expelling the Soviet occupiers and had opposed meddling in the country's internal affairs. He eschewed terrorism and the targeting of civilians. Taking jihad global, he warned, "could turn [jihadists] into bandits that might threaten people's security and would not let them live in peace."

On November 24, 1989, as the war in Afghanistan was drawing to a close, Azzam and his two eldest sons were killed by a car bomb as they drove to a mosque in Peshawar, Pakistan. Azzam's followers accuse Zawahiri of precipitating the final divorce between bin Laden and Azzam—by spreading rumors that Azzam was an American spy. Osama Rushdi, a leader of the Egyptian Islamic Group who knew bin Laden, Azzam, and Zawahiri, blames Zawahiri for Azzam's murder. Abdullah Anas, Azzam's son-in-law and a senior jihadist who fought in Afghanistan along his side, recalled that Azzam had complained bitterly to him about backbiting troublemakers, Zawahiri in particular, who spoke against the mujahedeen. In his memoirs, Anas reported that Azzam would say, "They have only one point, to create *fitna* ['sedition'] between me and these volunteers."

While we will never know who assassinated Azzam, two things are clear. First, bin Laden was the greatest beneficiary of Azzam's legacy. He took control of his Services Bureau and anointed himself "king" of the mujahedeen. He surrounded himself with Egyptian militants, theoreticians, and bureaucrats. Zawahiri and his lieutenant fell under bin Laden's spell, so that his hegemony over the Egyptian core of jihadism—al-shabab—was complete. Second, Azzam's absence from the Afghan scene meant that jihad would go global. The timing was perfect, for a year after Azzam's death Saddam Hussein invaded Kuwait and the American military intervention soon followed.

At first, Abu-Jandal and the Group of the North avoided bin Laden. "When Sheikh Osama heard there was an Arab group going to Tajikistan, he tried to contact us," Abu-Jandal remembered, "but we used to avoid him because we knew that if he got hold of us, we would not be able to get away. We knew bin Laden was in Afghanistan before we got there. Our avoidance was not from fear. We felt we would not be able to meet some of Osama bin Laden's requirements, although during our time in Afghanistan we did not know he had declared jihad against the United States and that he had decided to fight America. We did not know about that until after we returned from our journey north, that is, six months after we had headed toward Tajikistan." Abu-Jandal had no intention of staying in Afghanistan, which was merely a way station to his next destination, Chechnya. Chechnya declared its independence from Moscow in 1991, and since 1994, when the Russians invaded and tried to re-take control of the Chechen republic Ichkeria by force, the country has been through two wars. The estimated death toll of the first Chechen war (1994–1996) and the second (ongoing since 1999) ranges between 100,000 and 200,000 Chechen civilians and between 20,000 and 70,000 Russian troops. Almost 500,000 people have been displaced. Since September 11 Russian President Vladimir Putin has sought to portray Chechen fighters as Al Qaeda–like jihadists, and by so doing attract American support for the war in Chechnya.

Initially, no Islamist forces were fighting the Russians, but before long jihadists—mostly from Egypt, Algeria, and Saudi Arabia—gained prominence. Today it appears that they have taken control of the Chechen resistance from the more moderate and Sufi-oriented mainstream.

Abu-Jandal did not trust the Afghans after one of them betrayed his group in Tajikistan; some speculate that a member of Abu-Jandal's group, an ethnic Tajik Afghan, may have collaborated with the Tajik authorities against the Arab jihadists. The internal power struggle among the several Afghan factions also gave Abu-Jandal pause. Chechnya, he felt, would be better organized and would work along what he called "clearly identified lines." The youth was already becoming an adept strategist.

But the Egyptian al-shabab had not given up on recruiting Abu-Jandal; they knew he would prove a valuable asset and kept pressing him to meet with bin Laden. Abu Muhannad, a bin Laden associate, told Abu-Jandal over and over that Sheikh Osama had a cause to follow. " 'He has declared war on the United States,' " Abu-Jandal says abu Muhannad told him, " 'and we must fight the United States.' He then used to ask me to visit Sheikh Osama, and I used to say to him, 'This matter is not on my mind.' " Abu-Jandal continued to resist. " 'I do not want what bin Laden is calling me to,' " he told his friend. Pressed by another comrade, he answered, " 'My dear friend, let me be frank with you. I am not of the kind who wants to meet with Abu Abdullah [a term of endearment for bin Laden]. I am going to the front.' At the time," remembered Abu-Jandal, "we felt that staying in

Afghanistan without fighting would be a kind of torture and divine punishment and that God would be angry with us. We had dedicated ourselves to jihad, and the matter was finished. Our mission in life is to protect the ummah wherever we are able to go. If we heard that any young man from among our brothers had married, we used to go to him and offer him our condolences and not to congratulate him, as is the custom. We said to him: 'May God recompense you [an expression of condolence] for jihad.' And the matter was finished." Offering condolences was way of saying that anyone who opted for marriage had enlisted for servitude. Family obligations would then make it difficult for him to embark on the obligation of jihad. The context of this notion is the Qur'anic verse in which Allah says that one cannot be a true Muslim unless one holds Allah and his messenger Mohammed in higher esteem than one's loved ones and one's material possessions.

Eventually, Abu Mohammed al-Misri, a senior Al Qaeda leader and an Egyptian scholar, got Abu-Jandal to change his mind. "He said to me, 'Brother, it is the right of your Muslim brother that if he calls you, you respond to him. Responding to the call of Brother Abu Abdullah is something you owe.' I said to him, 'You *matawi'ah* ["enforcers of Islamic law"]! There is no escaping you!' Brother Abu Mohammed al-Misri laughed, and I agreed to go with him to Sheikh Osama." Later, Abu-Jandal discovered that bin Laden had sent for him.

Following the host etiquette of local tribesmen, bin Laden invited Abu-Jandal into his family quarters for three days. "During those three days, he waged a propaganda campaign in an attempt to convince us of the justification for his call for jihad against America. He told us about the bad state the Arabian Peninsula had reached and sought to convince us of the bad things that occurred there because of U.S. interference in the region." Bin Laden had tried to convince the ulema to take action but, he told Abu-Jandal, they were afraid to challenge the Saudi monarchy. Though Abu-Jandal's team knew in general terms about the American military presence in Saudi Arabia, they were not, as he put it, "aware of its hidden aspects," until he met bin Laden.

Bin Laden made a strong impression on the young jihadist, though doctrinal differences in the two men's religious perspectives—bin Laden's and Abu-Jandal's forms of Salafi were not the same, bin Laden's being more doctrinaire, more extreme, in orientation—left him with some doubt. "I did not fully accept Sheikh Osama's words, but they made me rethink. I began to verify some matters by contacting some people I knew and looking up books and publications published here and there, and in particular the reports and publications of the Saudi Reform Movement, or the Advice and Reform Committee (ARC), [Harakat al-Islah al-Saudi, which is hostile to the Saudi monarchy]. We began to contact one another regarding the case of Sheikh Salman al-Awdah, [a radical Saudi scholar who was jailed]. We began to [remember] the old tapes we used to hear in Saudi Arabia." Bearing such melodramatic titles as "From Behind Bars," "Steadfastness until Death," and "The Inevitability of Confrontation" these audiotapes contained sermons and

lectures by radical Saudi religious scholars who rose in opposition to the Saudi government in the wake of the first
Persian Gulf War. "After that we began to understand the
messages that those tapes conveyed to us in a manner that
we did not grasp in the past... We developed a stronger
awareness."

Abu-Jandal recalled that throughout the three days he
spent as bin Laden's guest, bin Laden hammered upon the
issue of the American occupation of the Arabian Peninsula.
"Of course, we were convinced there was a U.S. presence
and a U.S. occupation, but our view was different from his
view because of the fatwas... on the permissibility of calling
on the assistance of unbelievers." The then-Chief Mufti of
Saudi Arabia, Abd el-Aziz bin Baz, and his student, Mohammed bin Salih bin al-Uthaymayneen, had issued fatwas
authorizing the Saudi government to invite American troops
to help against the threat from Iraq. According to classical Islamic jurisprudence, Muslims are forbidden from seeking
help from non-Muslim military forces, or even allying with
them. Until these extraordinary and unprecedented fatwas,
this had been the default understanding among the Saudi
ulema and its people. Had these clerical leaders not sanctioned American military intervention in Saudi Arabia, Abu-
Jandal believed, there would have been an internal rebellion.
The religious sheikhs, however, were worried about something much worse, he said. "They wanted to ward off fitna.
Had fighting broken out inside Saudi Arabia between the invading U.S. forces that came under the guise of protection
and defense and the citizens of Saudi Arabia and the Gulf,

they would not have been equally matched in terms of resources and arms. The Americans are stronger militarily. The issue could have turned into a pretext that Washington could use to occupy the country. The young men were not prepared, trained, or qualified militarily for such an armed confrontation."

Initially, Abu-Jandal had accepted the logic of the fatwas in support of American troop deployment, since it was tied to the goal of expelling the Iraqi army from Kuwait and restoring that country's sovereign rights. He was therefore skeptical of bin Laden's argument. But by the end of the third day, he told *Al-Quds al-Arabi,* bin Laden had convinced him to join the fight against the Americans. He had felt pressured to come to a decision, for he was now not just part of the Group of the North but a member of its shura, and the others in the group were looking to him for guidance. "The young men began to ask me, 'What do you think, Abu-Jandal? What do you have to say on this matter?' I said to them: 'By God, every one of you is a man and should, therefore, be able to decide for himself.'"

Were it not for Abu-Jandal, the group might very well have departed bin Laden's camp with a polite thanks and farewell. But despite his admonition to the other members to think for themselves, Abu-Jandal's decision would determine their fate. By all accounts he had, as a young man, a quality, some would say an aura, that made him irresistible. Even the veteran *Al-Quds al-Arabi* reporter who interviewed him—nearly a decade after the period in question—seemed in awe of him: "He dazzles you with his arguments whenever you try to besiege him with questions," he wrote. Bin Laden knew that Abu-Jandal could sway his comrades. He had showered him with attention and praise during their time together. He always sat Abu-Jandal on his right side, a gesture of the highest respect. Whenever bin Laden spoke, he looked directly at Abu-Jandal. "Did you comprehend, Abu-Jandal? Did you understand, Abu-Jandal?"

Abu-Jandal's story of his recruitment is consistent with others' who fell under bin Laden's charm. (It is said that during bin Laden's second stay in Afghanistan, between 1996 and 2001, as many as forty thousand men received training at Al Qaeda's camps. Not all of them joined Al Qaeda, however. Many simply returned to their homelands.) The night before Ramadan, after afternoon prayers, bin Laden turned to him and asked, "'What is your view, Abu-Jandal, of what you heard?' I said to him. 'I will not hide from you, Sheikh, that what you said is convincing and that you are putting forward a clear case, but it is clear to me you do not have anyone from the people of the land itself, that is, from the people of the Arabian Peninsula, whose cause this is.'

"'What you say is true. Most of the brothers around me are Egyptians, Algerians, and North Africans,' bin Laden answered. 'That is why I invite you to join our caravan.' I had believed that Sheikh Osama had missed such a fact, but I discovered that he concentrated a great deal on the people of the Arabian Peninsula, especially on the people of Hijaz [the western province of Saudi Arabia]."

In fact, bin Laden long focused his recruiting campaign on Saudi Arabia, whose people share with him the Wahhabi view of Islam, though Al Qaeda represents an extreme off-shoot of the Wahhabi ideology. The Saudi regime, in turn, has waged a propaganda campaign against bin Laden and most of the kingdom's ulema have not sided with him. That is why bin Laden has focused on Hijaz. It is where Mecca and Medina are located, and its inhabitants tend to be both deeply religious and stubbornly independent.

After Abu-Jandal and his group joined Al Qaeda on the final night of Ramadan, "Sheikh Osama met with each of us separately, and many of us swore *baiya* ['allegiance'] to him immediately. Of course, the swearing of allegiance was very secret. No one knew who swore allegiance to him and who did not. Not all those who were around Sheikh Osama were necessarily members of Al Qaeda...Many young men had lived with him for two or three years and yet had not sworn allegiance to him and were not members of Al Qaeda, despite the fact they were constantly in his presence. Sometimes we used to hear that one of the young men had carried out a martyrdom operation [suicide bombing]. It was only then that we were sure that he had sworn allegiance to Al Qaeda. The execution of martyrdom operations enabled us to identify those who had sworn allegiance to Al Qaeda."

After swearing baiya, Abu-Jandal spent the next month in daily prayer and meditation. As he did with all the new recruits, bin Laden engaged him in "a form of mental mobilization." He was bent on molding Abu-Jandal's character through a rigorous regime of physical and ideological training. He had big plans in store for Abu-Jandal: He wanted to make use of the young man's rhetorical and administrative skills and turn him into a recruiter for Al Qaeda. Moreover, Abu-Jandal's Yemeni-Saudi background appealed to bin Laden. The bin Laden family had migrated from Yemen to Saudi Arabia, where they had amassed a fortune in construction. As I've suggested, tribal identity is seldom far below the surface in traditional and conservative Arab circles, even among the educated elite and professionals. Bin Laden trusted the Group of the North because they were mostly Saudis and Yemenis; their loyalty was thus almost guaranteed. They also offset the dominating influence of the Egyptians.

In his 2004 diaries entitled *The Story of the Arab Afghans,* which was published in the Arabic-language newspaper *Asharq al-Awsat,* Abu al-Walid al-Masri, a senior member of Al Qaeda's shura who had witnessed and participated in the network's most important decisions, confirmed the special relationship that existed between bin Laden and his followers. Everything revolved around bin Laden, Abu al-Walid recalled disapprovingly. Everyone wanted to tell bin Laden what he wanted to hear; he particularly captured the hearts and minds of the foot soldiers, who competed fiercely for his attention. The Saudi operatives worshipped bin Laden; he was their king, and they would do anything for him and for the cause.

A number of firsthand testimonies portray bin Laden as a cult figure, particularly among followers from the Arabian Peninsula, like Abu-Jandal. These young men from the Gulf found in bin Laden a heroic fatherly figure. Abu-Jandal reverently quoted Sheikh Azzam, who, before their schism, had urged loyalty to his protégé. "Osama bin Laden...is a whole nation embodied in one man. He shoulders the nation's cause."

Bin Laden's personal modesty and simplicity, his repudiation of the wealth and luxury into which he had been born, won the hearts of the young jihad recruits. "I never saw him pay attention to what he wore or to his personal belongings," Abu-Jandal recalled.

He used to sit with us and eat and drink with us. In the wake of the bombings of the U.S. embassies in Nairobi and Dar es Salaam in 1998, we were under economic siege and we experienced financial difficulties...Sheikh Osama came to visit us and asked: "What are you eating today?"

"Rice and potatoes," I replied.

"May God be praised, you are richer than we are," he answered.

"Richer than you, how?"

"We ate bread and yoghurt today." He used to lead a very simple life.

I remember that I told him, "Sheikh, the situation here is difficult for the young men. I am here in charge of forty Arabs and Afghans and we have no money or anything to eat." He laughed and said, "My son, Jandal, we have not yet reached a condition like that of the Prophet's companions who placed stones against their middles and tightened them around their waists. The messenger of Allah used two stones."

I looked at him and said, "Abu Abdullah, those were men who had strong faith and God tested them, but we are weak and God will not test us in the same way."

He looked at me in astonishment and said, "Are you really saying these words?"

I replied, "Yes, those men were strong in faith and God wished to test them. We, on the other hand, have sinned and God would not test us." Sheikh Osama laughed.

Imagine a man with the kind of resources he had, the cause he embraced, and his stature as a leader—sitting with us and eating rice and potatoes. I remember that at one point we ate only dried bread and water. Sheikh Osama used to take hard bread, dip it in the water and eat it, saying: "May God be praised. We are eating but there are millions of others who wish they could have something like this to eat." So we never really felt afraid as long as we were with that man.

Life within the bin Laden household was apparently equally austere. Though Abu-Jandal was never introduced to the great leader's daughters—it would not have been customary—he alternated doing guard duty with the older bin Laden sons. "I used to hear him telling his sons, 'Your father's millions about which you hear are not for your father to use. This money is for the Muslims and I hold it as a trust for the cause of God. Not one riyal of it is for you. Each of you is a man. Let him rely on himself.' His sons actually worked and each developed an income of his own with his sweat. This was the kind of life to which Sheikh Osama accustomed his sons."

Except, that is, for his eldest son, Abdullah, who broke with his father over the primitive conditions and returned to Saudi Arabia. Of bin Laden's four wives, one, Umm Ali, asked for a divorce while they were living in Sudan. Bin Laden granted the divorce. His three other wives, all from wealthy, educated families, remained with him in Afghanistan. Aside from two children by his former wife, his sons and daughters also lived with him in Afghanistan.

For recreation, the bin Ladens, accompanied by Abu-Jandal, went on day trips into the wilderness outside Kandahar. "Bin Laden would ride with us in his private car and his family would ride in a bus. His grown sons would follow us on horseback, though the distance was about an hour by car. The sheikh would then sit with his wives and we used to take the car far from that place so that he would not see us and we would not see him. We would communicate by radio. Bin Laden would teach his wives how to use firearms. They would play together and do some simple physical exercises; bin Laden led a normal family life."

Joining Al Qaeda meant joining the war against the United States. Abu-Jandal boasted that America would not be able to defeat them. "In view of our military experience and our experience in carrying arms, we said: What is America? If we had succeeded in many armed confrontations and military fronts against the Serbs, the Russians, and others,

America would not be anything new. We often sat down with the brothers who fought the Americans in Somalia, and we used to hear about the brothers who struck the Americans at the Aden Hotel [in Yemen] in the early 1990s and about the brothers who blew up American residences in Riyadh and al-Khobar."

There had been two terrorist attacks in the Saudi kingdom, the first the truck bombing of a U.S.-operated Saudi National Guard training center in Riyadh in November 1995, in which five Americans and two Indians were killed; the second the bombing of the Khobar Towers housing U.S. Air Force personnel in June 1996 in the eastern city of Dhahrah. Nineteen servicemen died and another 372 were injured. "We reached the conclusion that America was no different from the forces we had fought because it had become a target for all and sundry. All of its foes had dealt blows to it… That was the beginning of my work with Al Qaeda.

"We had trained in Bosnia and gained some practical experience in Somalia and Tajikistan," Abu-Jandal noted, but they developed real fighting skills in bin Laden's training camps in Afghanistan. "I participated in many military courses, such as a tactical foundation course in guerrilla warfare and some special courses in the use of automatic weapons. Whenever I completed a course, Sheikh Osama used to summon me, and I used to go and stay with him."

Abu-Jandal continued to grow in bin Laden's estimation. After the young man reacted quickly to an apparent assassination threat, he was made bin Laden's personal protector. (There were several assassination attempts on bin Laden's life. According to State Department documents declassified on August 18, 2005, by the National Security Archive under the Freedom of Information Act, top American and Taliban officials discussed eliminating Osama bin Laden after the bombings of the U.S. embassies in 1998. Alan Eastham Jr., deputy chief of mission at the American Embassy in Pakistan, met with Wakil Ahmed, a close aide to Taliban leader Mullah Omar, on November 28, 1998, at the Taliban embassy in Islamabad. In that meeting, Wakil Ahmed told Eastham that the Taliban understood the U.S. desire to have bin Laden expelled from Afghanistan. One way would be if the U.S. were to "kill him, or arrange for bin Laden to be assassinated." According to Wakil, the Taliban wanted to "resolve" the "bin Laden problem" as quickly as possible because failure to do so would result in American bombing of Afghanistan and the Taliban's "termination.")

"After that day Sheikh Osama gave me a pistol and made me his personal bodyguard," Abu-Jandal said. "The pistol had only two bullets, for me to kill Sheikh Osama with in case we were surrounded or he was about to fall into the enemy's hands, so that he would not be caught alive. I was the only member of his bodyguard who was given this authority. I took care to keep the two bullets in good condition and cleaned them every night, while telling myself, 'These are Sheikh's Osama's bullets. I pray to God not to let me use them.'

"I constantly stood behind his back and accompanied him round the clock. I went with him everywhere..." It wasn't that bin Laden was afraid. He just took precautions.

"If something happened to him, the nation would suffer moral and psychological defeat. However, if he refused to be captured alive, he would become a martyr, not a captive, and his blood would become a beacon that arouses the zeal and determination of his followers. I remember that he used to say: 'Martyrdom rather than captivity.'"

Relations inside the bin Laden compound were not always harmonious, especially among the clique surrounding the leader. As bin Laden's personal bodyguard, Abu-Jandal's duty was to defend bin Laden. Sometimes, however, the Sheikh sent him as his representative on missions, such as to resolve disputes among Al Qaeda factions.

Like all bureaucracies, Al Qaeda was beset by internal quarrels and rivalries; its members were not above the fray of material and political self-interest. Evidence obtained from captured Al Qaeda computers in Kabul immediately following the fall of the city reveals infighting and jealousies among various jihadist factions, stemming from favoritism and penny-pinching. The case of a Sudanese named Jamal Ahmad al-Fadl is illustrative. Before his defection in 1996, al-Fadl was a trusted operative in the bin Laden network, a money carrier dispatched on sensitive missions to Pakistan, Sudan, the United States, and elsewhere. He told American officials that he felt exploited by bin Laden, that he had not been compensated fairly, that others received much higher salaries than his own. Accordingly, he scammed tens of thousands of dollars from Al Qaeda, and when caught and ordered to return the stolen funds, he walked into a U.S. embassy and offered his services. He is now in a witness protection program.

"There were rivalries among Al Qaeda members depending on their countries of origin," confirmed Abu-Jandal. "The Egyptians used to boast about being Egyptian," he said. "The Saudis, Yemenis, Sudanese, and Arab Maghreb citizens used to do the same thing sometimes. This troubled Sheikh Osama and he used to send me to help eliminate these regional rivalries because the enemies of God, those who have sickness in their hearts, and informants would exploit these ignorant attitudes and try to sow divisions and disagreements among Al Qaeda members." But the antagonism persisted, especially among those Al Qaeda members who felt that the Egyptians dominated the network and controlled most of the important committees. Al-Fadl testified, in English, in a U.S. court about how he and other disgruntled non-Egyptians had once confronted bin Laden and his top aides, Abu Ubaidah al-Banshiri and Abu Hafs al-Masri. "We told Abu Ubaidah and Abu Hafs and bin Laden in that meeting that the camp run by Egyptian people and the...emir from the guesthouse is Egyptian and everything Egyptian people and form the Jihad group, and we have people from Nigeria, from Tunisia, from Siberia, why is Egyptian people got more chance than other people run everything? I tell them the people complain about that, but the people embarrassing to tell them, to tell you face-to-face, but most of the people, they complain about that." According to Al-Fadl, bin Laden brushed such complaints aside, gently but firmly reminding his foot soldiers that they were all Muslims serving "the cause of God."

However, the warriors of God had unconsciously internalized Western nationalism and behaved more like citizens of separate tribes than a unified ummah. Abu-Jandal suggested that bin Laden eventually absorbed the criticism that Egyptians dominated his inner circle and labored hard to rectify the ethnic imbalance by recruiting Saudis, Yemenis, and North Africans. By 2001 "the biggest percentage" of "ordinary members" was Saudi–Yemeni, like Abu-Jandal and most of the September 11 suicide bombers.

Abu-Jandal recounted a dramatic encounter when a senior Taliban delegation headed by Prime Minister Mullah Mohammed Rabbani publicly challenged bin Laden to prove that he had followers from the Arabian Peninsula.

They expressed a wish to be introduced to all those who were present. The brothers began introducing themselves: I am so and so from Egypt, I am so and so from Algeria, I am an Egyptian and my name is, I am from Iraq, I am from Morocco. Mullah Rabbani looked at bin Laden and said: "Sheikh Osama, we have not heard anyone say I am from the Gulf. The Saudis have said that you do not have with you anyone from your country, the people of the cause. They say all those around you are a group of Egyptians, not from Saudi Arabia or the Arabian Peninsula." Sheikh Osama was provoked by this comment and said: "What do they mean I have no people from the Arabian Peninsula?" He turned to one of our young companions who was standing by the door and said, "Call our Arab brothers!" More than seventy men showed up.

Sheikh Osama began introducing them: So and so from Mecca, so and so from Yemen, so and so from Jeddah, this man from Riyadh, et cetera. The situation completely changed and the Egyptian brothers felt lost among so many people from the Arabian Peninsula. Mullah Rabbani's eyes filled with tears and he said: "All these are people from the land of the two holy mosques." He assured Sheikh Osama that the Taliban would always defend him and his followers despite the pressure that Saudi Arabia was putting on the Taliban movement. He told him: "As long as all these men are with you, then you have men and do not need us. Still we consider ourselves your allies." These words were very moving to Sheikh Osama and most of the young men who were present. This was one occasion when I saw him show emotion. He was moved not because of personal pride or out of a wish to be the leader of a cause merely in name. No, the cause he embraced had become a general cause and many people joined him in defending it. It was not Osama bin Laden's cause but had become the cause of many people around him in that region.

Abu-Jandal returned to Yemen from Afghanistan in the summer of 2000, about two months before a small boat packed with between 500 and 700 pounds of explosives rammed the USS *Cole* in the port of Aden. Yemeni authorities arrested Abu-Jandal after that and imprisoned him for

twenty months; he spent thirteen of them in solitary confinement. Six other men, all veterans of the Afghan war, were also
arrested in connection with the attack on the USS *Cole*. Bin
Laden approved and funded the attack. Now Abu-Jandal
lives and runs a private water-processing company in Sana'a.
In an interview with the BBC in the fall of 2005, he said that
while he himself had given up armed struggle, he remained
more convinced than ever about the justice of Osama bin
Laden's cause.

IV

The GREAT SATAN,

NEAR AND FAR

THE GREAT SATAN. No single phrase has jangled in the American ear more than the one delivered by the craggy-faced, white-bearded leader of the Iranian Revolution, the Ayatollah Khomeini. "Bloodthirsty warmongers" or "neo-imperialists"—these were sobriquets Americans had grown used to hearing during postcolonial skirmishes when Americans were sometimes told to go home. But what to make of a relatively modern and reasonably educated population like that in Iran wrapping up its anti-Americanism in the black robes of the clergy? Americans, who spent—then as now—more hours in church worship than the inhabitants of any other developed nation, whose president had proudly proclaimed himself a born-again Christian, seemed utterly baffled by being linked to the forces of Satan. Surely this band of angry students and radical mullahs was nothing more than a demented anomaly, many Americans felt, soon to pass, like other messianic communalists.

Except things didn't turn out that way. The image of America as the benign beacon of hope to a hungry world was rapidly disintegrating in the minds of millions—perhaps even tens of millions—of Muslims, and non-Muslims, who were coming to see the Stars and Stripes as the manifest face of evil on earth. Unlike those from Marxist and socialist critics, the Islamist attack on the United States did not focus only on class division, racism, and injustice. Rather, it rejected the basic idea of what America at its best represents. It was a spiritual critique every bit as profound as Galileo's conflict with the Inquisition over the position of the sun in the universe. If we do not bring ourselves to understand the moral and spiritual mission that drives today's jihadists, we will remain forever baffled at the "irrationality" of their speech and the "insanity" of their actions.

Even as someone who still feels more at home in the narrow market streets of the Arab soukh than in the vast spaces of Macy's or Nordstrom, I, too, was baffled by the Satanic characterization of America. What could possibly have inspired such widespread loathing for a nation that had invited Muslims, by the tens of thousands, into its universities? Do they confuse America with its foreign policy? Or do we equate their denunciation of the American ruling elite with a denunciation of the American people? What fuels their anger and rage against American power?

When I posed these questions to Kamal, he advised that I return to the writing and speeches of the martyred Sayyid

Qutb, whose two years of study in America, from 1948 to 1950, had so shaken him that he became convinced that the American dream was nothing less than a Satanic plot against the Kingdom of God. More than any other modern document, Qutb's "The America I Have Seen" continues to provide ideological nourishment to Islamists and jihadists, just as it has since it first appeared in 1950.

On the ship that carried him across the ocean to America, Qutb experienced a formative event that convinced him of the general moral decadence of the West. One night after he had gone to bed, he heard a soft knock on the door of his cabin. Opening the door, he found a tall, "stunningly beautiful" woman, scantily clad, her body "sensuous" and "inviting."

"'Will you permit me to be your guest, sir?' she asked.

I apologized because it was a single room with a single bed.

'A single bed often accommodates two people!' she answered."

Qutb recounted how he had to physically prevent the woman—whom he believed to be drunk—from entering his room, closing the door in her face, leaving her to "stumble and fall in the hallway." It was a bad omen for what awaited this innocent schoolteacher. America, he would write, was a land filled with "pleasures that acknowledge no limit or moral restraint, dreams that are capable of taking corporal shape in the realm of time and space." Qutb was neither the first nor the last visitor to the New World to be struck initially by what they saw as moral depravity. In the mid-twentieth century, traditionalists of many stripes were shocked by what was happening in postwar America—the overwhelming numbers of new cars, the housing developments springing up everywhere and taking over small-town America. Sons and daughters were abandoning their family homes, moving sometimes thousands of miles away, and visiting their parents one or twice a year. As Hollywood's moral codes loosened, James Dean and Rock Hudson swaggered bare-chested across the screens of movie houses. Scantily clad women sold everything from Chevys to industrial drill bits. For Americans, liberated from nearly two decades of war and depression, freedom was in the air. For someone like Qutb, America represented a gigantic bright red For Sale sign. Everything was for sale, including the soul itself.

"Rotten and ill" to its core, Qutb wrote. It is a judgment that shaped, and continues to shape, the jihadist view of America—and of the West in general. On the surface, the denunciation might not seem all that different from the Marxist critique of capitalism, in which all that glitters turns to dust, or even from the Puritan view of the body as a putrid vessel of iniquity. Yet the Islamist's outlook is neither Puritan nor socialist, however. Market exchange forms the basis of Islamic community life and no poetic tradition contains verses more lush or sensuous than those of the Ottoman Hüsrev to his beloved Shirin. What perplexes so many Westerners, as I have often heard, is how such an outwardly austere and sometimes bloody culture like the Muslim world could nurture tenderness that is at once so deeply sensual and so profoundly spiritual.

As I've suggested, the most basic differences involve time and space. In America and in most of the Christian West, religious feeling normally gets expressed in restricted and often private circumstances: inside the church, at the dining table, in the confessional. For most Christians, common worship only takes place on Sundays. Muslim cultures view religious expression as a public, communal act, not confined to one time slot a week; it is a daily journey (one of the five pillars of Islam is prayer five times a day, at sunrise, noon, afternoon, sunset, and evening). In warmer climates the heart of the mosque is open to the air, and the very act of removing one's shoes before entering the sacred space places all worshipers in a common collective status. And because collective prayer occurs daily, it requires believers to be continuously mindful of God's presence. The first *adhan*, or call to prayer, summons Muslims for the dawn prayers and marks the beginning of a new day, a new gift from God. Christian Arabs are as imbued with this collective sense of spirituality as their Muslim counterparts are. The mosques are most crowded on Fridays, churches on Sundays. (Making Lord Cromer, the powerful British administrator in Egypt during the height of Victorian rule, lament that the only difference between Muslims and Copts (a Christian sect) in Egypt was that the former went to the mosque and the latter to the church.) The average Arab constantly senses the presence of God, to whom is due the new day and all things. Tribulations must be endured because they, too, are from God and he is fair and most compassionate. Therefore, things will have to change for the better.

A typical day in the alleys of Cairo or Beirut begins with shopkeepers loudly praising God and invoking him to bring good fortune—a ritual that of course simultaneously draws attention to their wares. The poor and indigent circulate past the shops with incense, receiving in exchange a small charity—a blessing that increases the likelihood of a good day. Unlike in the West, there are no lonely crowds in the Muslim world; crowds are social occasions. People wake up in the morning and drink sweet, thick Turkish coffee with their relatives or neighbors. Some breakfast on sweets called *knafe*, pastries composed of cheese, ground wheat, and rose water. And it's common to bring a few cakes with you to share with coworkers when you get to the office. Generosity in the Middle East is in the air people breathe. Here thrives the cult of the guest. Should someone invite you to lunch in an Arab country, it won't be a simple affair. You will be offered a banquet whose appetizers alone would be enough for most Westerners. Arabs like to say that the table should satisfy the eye before the stomach. Western visitors may be put off by such excess, but Arabs take pride in being generous and welcoming.

Evenings are longer than the days in the Arab world. Relatives and friends routinely drop by unannounced. In many family homes, for example, as many as twenty friends and relatives might drop in every evening and stay well past midnight. And it would be unseemly if one did not try to convince a guest to stay still longer. The idea of community is not taken for granted, nor is it a matter for debate and discussion. It is a living reality practiced on a daily basis and perceived as an expression of spiritual health. Even large cities like Beirut and Cairo, the most Westernized Arab capitals, have the feel of small towns. The Lebanese poet Shawki Bazi describes Beirut as a cluster of villages. Writers, professors, and artists crowd sidewalk coffee shops, but unlike their counterparts in Paris, Berlin, or New York, they are usually not total strangers to one another.

Contrasting this pervasive sense of camaraderie and instinctive magnanimity, Sayyid Qutb found in America an alluring but ultimately vapid landscape. At first this Sunni Arab was fascinated, even seduced, by the inherent optimism of America's rich polyglot heritage. "Imagination and dreams glimmer in this world of illusion and wonder. The hearts of men fall upon it from every valley, men from every race and color, every walk of life, and every sect and creed... America is the land of inexhaustible material resources, strength, and manpower. It is the land of huge factories, unequaled in all of civilization... American genius in management and organization evokes wonder and admiration. America's bounty and prosperity evoke the dreams of the Promised Land."

Yet no sooner had he cataloged the glories of the American cornucopia than he began to sense rot from within: "It is the case of a people who have reached the peak of growth and elevation in the world of science and productivity, while remaining abysmally primitive in the world of the senses, feelings, and behavior. A people that has not exceeded the most primordial levels of existence, and indeed, remains far below them in certain areas of *feelings and behavior"* (emphasis added). While in America Qutb certainly understood the allure of technological progress but, for him, materialism could not obscure the moral and intellectual ends of society. "Humanism," he later wrote in his book *Islam: The Religion of the Future,* "is the ultimate scale by which man's development or his backwardness should be measured, while his spiritual happiness is the measurement of the fitness or unfitness of the elements of his civilization to his nature."

For Qutb, whose youthful ambition was to be a poet, the engine of Western progress had become so all-consuming that it destroyed everything in its path, annihilating the very terms of intimacy that define what it means to be human. He and his contemporary followers are not alone in that perception, as we can see in today's growing antiglobalization movement. When the machinery of productivity obliterates all other human values, he argued, the very basis of morality is lost. The failure of America's moral impulse derives, Qutb theorized, from the peculiar nature of its creation, the result of its "deformed birth"—a crossing of Enlightenment European science with the New World's wild, untamed nature. "In America," he wrote, "man was born with science, and thus believed in it alone...Since he received nature as an untamed, stubborn virgin, and fought to build his homeland with his bare hands, applied science was his greatest ally in his violent struggle. Applied science reached out to him with effective tools for creating, building, organizing, and producing."

The American's obsession with technology came at a high cost to his heart and soul, believed Qutb. Unlike the brilliant late-medieval Arab and Muslim scientists, who never had to set themselves in opposition to religion, the American's denatured scientific quest "narrowed his horizons, shrank his soul, limited his feelings, and decreased his place at the global feast, which is so full of patterns and colors." Or, as Qutb lamented in plainer language in a letter to the Egyptian playwright Toufic al-Hakim, "I wish I could find somebody to talk with about human affairs, morality and spirit—not just dollars, movie stars, and cars."

We can find echoes of Qutb's disenchantment with a commerce-driven America in the observations of many other visitors to the United States over the years, but his near-obsession with American sexuality and the independence of American women sets his apart. Then as now, female sexuality was seen to threaten the very core of jihadists' authority. The public display of women's bodies demonstrated to Qutb an animalistic "primitiveness" that belied all other claims of progress. The Egyptian could simply not stomach the sexual license of the American man and woman. "The word 'bashful' has become a dirty, disparaging word in America," he wrote. Based upon what he experienced at Stanford, he wrote,

> *Some philosophize about it, such as the girls in the university, who once told me: "The matter of sex is not a moral matter at all. It is but a question of biology, and when we look at from this angle it becomes clear that the use of*

words like moral and immoral, good and bad, are irrele-
vant." It may appear that Americans are not only strange,
but amusing. Some of them excuse themselves and justify
it as one doctoral student did: "We here are occupied with
work, and we do not wish to be hindered from it, and we
do not have time to invest in feelings. Moreover, books try
our nerves, so we wish to do away with this worry to free
ourselves for work with relaxed nerves!"

It might be tempting for many Western readers to dis-
miss Qutb's views as warped and one-sided. But doing so
would be to ignore the expansiveness and nuance of his
writings. Qutb saw the family unit as key to the development of progressive political and social values. The difficulties facing the Civil Rights movement and the jingoistic
rhetoric of the McCarthy era in America convinced him that
racism and nationalistic fervor, absolutely contrary to the
Islamist program, were results of the degenerating social
fabric. According to Qutb, a society that wanted to combat
those forms of discriminations and their primitive origins,
"while providing full opportunities for the development
and perfection of human characteristics, requires strong
safeguards for the peace and stability of the family." Any society in which "immoral teachings and poisonous intentions
are rampant," and sex is considered "outside the sphere of
morality," the "humanity of man can hardly find a place to
develop."

That said, seemingly unaware that the permissive sexual
mores he found in California would have been quite differ-

ent from those in the American heartland, Qutb felt lost in a wanton sexual jungle. "Body to body, and female to male. On the basis of bodily needs and motives, relationships are based and ties are established. And from them stretch the rules of behavior, the mores of society, and the ties of families and individuals...With the temptation of the body alone...girls meet boys, and from the strength of the body and its muscles the boy obtains the submission of the girls. And the husband obtains his sexual rights, and those rights disappear completely the day the husband fails to 'perform' for one reason or another."

To be a single, lonely, forty-four-year-old foreigner, walking across the sun-dappled Stanford campus, where young men and women strolled arm-in-arm together, was too much. The "American temptress" preyed constantly on his mind.

The American girl is well acquainted with her body's seductive capacity. She knows it lies in the face, and in expressive eyes, and thirsty lips. She knows seductiveness lies in round breast, the full buttocks, and in the shapely thighs, sleek legs and she knows all this and does not hide it. She knows it lies in clothes: in bright colors that awaken primal sensations, and in designs that reveal the temptations of the body—and in American girls these are sometimes live, screaming temptations! Then she adds to all this the fetching laugh, the naked looks, and the bold moves, and she does not ignore this for one moment or forget it!

As he wrote after his return to Egypt in the magazine *Al-Kateb,* the innocent American girl turns into a diabolical vixen: "[A]s soon as she gets closer to you, you feel her overpowering sexual drive devoid of any innocence, then she turns into flesh, mere flesh, but nonetheless real voluptuous flesh."

The American "dream boy" fares little better, in Qutb's eyes. Such boys know full well that the wide, strapping chest is the lure that attracts girls who fall for "cowboys." A young nurse in a hospital told him frankly, "I want nothing in the man of my dreams but two strong arms he can really squeeze me with!" He found a *Look* magazine survey of several girls of varying ages and education in which "the overwhelming majority declared their open attraction for boys with the so-called ox muscles."

Qutb may have written this over a half a century ago, and though he might seem to Americans as both comically naive and deeply repressed, he has more followers in the Arab world today than he did during his lifetime. *The America I Have Seen* was reprinted in 1993 by an Islamist publishing house in Egypt and it continues to sell briskly. In the foreword of the new edition, activist Mohammed Abed al-Mun'im reminds readers that Qutb's words are as relevant now as they were in 1950. On the cover is a portrait of Qutb; behind him, a deformed Statue of Liberty is pictured standing on a pile of skulls, her tablet transformed into a short-fused bomb, her torch reconfigured as an incendiary device. Mamdouh Ismail, a radical Islamist who spent three years in prison for his alleged role in the Sadat assassination, told me that Qutb "set my world on fire. [He] taught me that our Islamic nation is morally superior to the West and the United States. We must not be deceived by American and Western propaganda about progress and civilization. We must not let America colonize us mentally. We are facing a brutal form of intellectual and ideological colonization."

Nowhere does that cultural colonization strike harder at the Islamic ideal than in the ubiquitous and popular American media. As never before, Hollywood has now brought licentiousness to Muslim homes in isolated villages and hamlets through the sale of videotapes and DVDs. In the eyes of jihadists, they have undermined religious and parental authority over the behavior of adolescents and adults alike. "Morality in America is long dead," a young Lebanese militant named Hashem told me. I asked why, hoping for details. The answer he gave was too much mixing of the sexes and too much sexual freedom: "Look at how girls dress, go out alone with boys and conduct themselves on the streets," he said. "They are easily exploited." Again and again, jihadists insisted—with defensive emphasis, I thought—that America and the West were in no position to lecture them about women's rights and freedoms.

This attitude is shared by more than just religious radicals. When I was in Cairo I went into the old city to find Gamal al-Banna, the brother of Hassan al-Banna, the founder of the Muslim Brotherhood and a leading authority on Islamic texts who also happens to be a forceful critic of the conservative religious establishment. Al-Banna was seated behind a desk crowded with books while we talked. What, I asked him, did he think about the empowerment of women, including the idea of outlawing the Muslim man's traditional privilege of marrying up to four wives? "Multiple wives or multiple mistresses!" he exclaimed.

"What do you mean?"

"Look," he answered, "everyone in America cheats on their spouse. Every American male engages in sex before marriage and continues to do so outside of his marriage. This is common knowledge. In contrast, Islam regulates and sanctions sexual needs and desires within a marriage. If a Muslim man needs more than one woman, he can do so but he must sanction the relationship through marriage. Tell me, which is a more rational and just system? Which of the two systems respects better the dignity and rights of women— the Islamic or the American?"

Even worse than Hollywood, in the eyes of Islamists, are American churches, institutions supposedly created for prayer to the same god worshipped by faithful Muslims. But churches, thought Qutb, have been degraded and desecrated by social corruption, the kind that sends innocent young boys into the waiting arms of fallen temptresses. During the two years he spent in America, Qutb made a point of visiting local churches wherever he went. American churches, he railed, had become nothing more than sites for "carousal and enjoyment or, as they call it in their language 'fun.'" More shockingly, the ministers approached religion like a business, competing with one another to attract larger congregations, performing before the altar like "merchants or showmen or actors." In a chapter entitled "A Hot Night at the Church," Qutb recounts an evening he spent at a church in Greeley, Colorado. "I was a member in its club as I was a member in a number of church clubs in every area that I had lived in, for this is an important facet of American society, deserving close study from the inside." After the religious service, everyone proceeded to the dance floor and every boy took the hand of a girl. "The dance floor was lit with red and yellow and blue lights, and with a few white lamps. They danced to the tunes of the gramophone, and the dance floor was replete with tapping feet, enticing legs, arms wrapped around waists, lips pressed to lips, and chests pressed to chests. The atmosphere was full of desire."

Then, still more shocking (as it would have been to twentieth-century Methodists as well), the minister stepped out of his office, looked intently at the dance floor, and encouraged those who had not yet participated in this "circus" to dance. Noticing that the white lamps spoiled the "romantic, dreamy" atmosphere, he dimmed them one by one until the place really became more "romantic." Finally, Qutb tells us, the minister advanced to the gramophone to play a song that would befit this ambiance and chose a famous tune called "Baby, It's Cold Outside," in which a boy invites a girl to his home and as the evening grows late entreats her not to go home because it's cold outside.

Despite its inherent opposition to the Islamist program, Qutb thought the West was a deeply spiritual society at odds with itself, determinedly far removed from secular and materialistic Soviet communism. Liberal capitalist thought, though suspicious of religion, never eliminated spirituality from society; religion was banished to the strictly private realms of ritual and routine, where it developed an agonizing tension with the increasingly atheistic public sphere. Qutb keenly picked up on this "hideous schizophrenia" in America, pitting its Protestant heritage irreconcilably against its modern scientific skepticism: To support his point Qutb cited the late Secretary of State John Foster Dulles, who, in his book *War or Peace,* lamented the spiritual poverty in American life and called for reclaiming and deploying faith in the struggle against world communism: "Something has gone wrong with our nation...That is new in our history. The trouble is not material...What we lack is a righteous and dynamic faith."

I was curious what a worldly, moderate Islamic scholar like Gamal al-Banna would have to say about Qutb's depiction of religion and church life in America, which militant activists had adopted wholesale. As I sat in his study, he expounded on the point where he believed Islamic and Western spirituality diverged.

America is an extension of Western culture, he began, one that worships "power and freedom, not religion and God. There is a fundamental, philosophical divide between our Islamic civilization and that of the West. American and Western culture," he went on, "elevates the freedom of man to a sacred status—man has replaced God as the focal point of Western civilization. Individual happiness comes before everything else, including religion and morality. Not for us Muslims. God is the center of our life and existence." That was why Qutb found fault with almost every facet of the

American way of life—religion, character, art, sex, food, clothes, and hairstyles. Yet Qutb was no Tocqueville. He hardly touched the surface of American society. How can we therefore take Qutb's diatribe against America and Americans seriously?

Mamdouh Ismail, the attorney who defends jihadists, told me that unlike secular Muslims he and others do not look up to the West and America as models. "We feel free of American hegemony because we believe in the ultimate truth of the Qur'an and Shariah. The fifth column in our midst failed to sell the West and America to our people." Even Gamal al-Banna expressed doubts that the West had anything to offer Muslims socially and morally. "The West and America are desperately in need of values to preserve their civilizations," he said. "While America and Europe have freedom, they no longer believe in timeless values and truths. Islam has values but no political freedom. Our challenge is to balance values and freedom. The challenge facing America and Europe is to reclaim their moral compass, which cannot be had without religion." And yet as I spoke to all these Muslim men—speaking to observant Muslim women is very difficult—I still found myself puzzled by the depth of cultural misunderstanding of American society.

To find answers, I was advised go back and read Qutb, who had turned against America even before his ship had set sail for New York. The reason wasn't cultural but political. It was because President Harry Truman had actively and effectively lobbied the United Nations for the partitioning of Palestine into two states, one Jewish and the other Arab. The

Truman administration was among the first to recognize the establishment of Israel. That act, according to Qutb, marked the end of innocence about America in the eyes of Muslims; they learned that the shining city on a hill was merely another colonial power, willing to trample their rights to advance its global interests. Qutb published an article in *Al-Risala* in October 1946 called "The American Conscience and the Palestine Question," in which he railed against America's "treachery" and "duplicity." This important document, which has never been translated into English or much examined by those in the West, became a philosophical cornerstone of Islamist hostility toward Israel and its Western allies.

> *We finally discovered the U.S. conscience that had captured the hearts of many people in the East, who considered it to be different from the British conscience and the French conscience and those of the rest of Europe...Many had been deceived by the American conscience because they had less contact with America than with Britain, France, and Holland. But America's role in Palestine exposed the deceptiveness of the American conscience that gambles away the future of other people and their human rights to purchase a token of votes in the presidential elections...This is America exposed for all to see. This is Truman revealing the truth about the American conscience, which is the same as every Western conscience—unscrupulous, and only fools trust it.*

Qutb made no effort to conceal his disdain of Western politicians: "I cannot convey my hatred and contempt for those Westerners!" Even worse, however, were "those Egyptians and Arabs who still put their faith in the Western conscience generally and the colonial conscience specifically." He challenged Muslims to prepare for the coming struggle: "If you want to be saved from the jaws of the Western beast, there exists only one way out...: begin the jihad and ignore any traitor who tries to trick you into trusting the Western conscience." All Arabs, all Muslims, needed to stand up to defend Palestine. "It is the struggle between the rising East and barbaric West—between God's laws and the laws of the jungle."

It would be convenient to believe that only the extremists followed Qutb's lead in their hostility toward Israel and its allies. The truth is that Arabs—not just Islamists—thought that American politicians had stabbed the Palestinians in the back, sacrificing them to placate the powerful Jewish community believed to dominate American political life. The Israeli-Palestinian conflict has shaped the perception of Arabs and Muslims toward America far more than anything else. On this score there exist no differences between Islamists and secularists, leftists and conservatives; all blame America for tipping the balance in favor of the Jewish state.

Not surprisingly, every Islamist and jihadist I have ever interviewed has made a point of condemning America's policies toward Israel. In his "Letter to America" published in the British *Observer* on November 24, 2002, after being posted on Al Qaeda's Web site on October 14, 2002, bin Laden tried to explain to Americans why he had launched his attacks on them. Palestine topped his list of grievances: "The creation and continuation of Israel is one of the greatest crimes, and you are the leaders of its criminals...It brings us both laughter and tears to see that you have not yet tired of repeating your fabricated lies that the Jews have a historical right to Palestine, as it was promised to them in the Torah." Bin Laden's opinion is widely shared by those who denounce his violent methods. "The West and America planted the Zionist entity in the heart of the Arab world to keep it weak and divided," Mamdouh Ismail told me. "The West never really ended its imperial presence in the region because it left Israel behind as an armed fortress to do its bidding. Israel is an extension of the West and America's colonialism; it could not and cannot exist without U.S. military, financial, and political backing."

I asked Gamal al-Banna about his views on Israel, America, and the Arabs. "The creation of Israel," he told me, "was one of the most important reasons Arab-American relations were poisoned. It led to the estrangement of Arabs from America. It also gave a boost to the Islamic movement. Arabs and Muslims have never accepted Israel as a legitimate state. They view it as an American base in the Middle East. Active U.S. support for Israel against Arab countries in the 1967 and 1973 wars reinforced widely held perceptions of American hostility."

Al-Banna reminded me that before Qutb, his brother Hassan had been among the first Islamist leaders to warn Muslims about what he saw as a Western plot involving Israel. Hassan had been expounding that message when he was assassinated. "When words are banned," Hassan had declared after the Muslim Brotherhood was banned in 1948, "hands make their move." His death marked the beginning of the civil war between Islamists and the Egyptian government.

Qutb believed that the Palestinian issue encapsulated a clash of civilizations, if not religions. The ultimate showdown would not be between the great superpowers of the Cold War, but, as he wrote in his 1949 book, *Social Justice in Islam*, "between Islam, on the one hand, and the Western and Eastern blocs, on the other." Islam was the only force resisting the power of Europe and America. "The real struggle in the future will not pit capitalism against Communism, or the Eastern camp against the Western camp...it will be between materialism throughout the world and Islam." A conspiracy was being hatched between the Eastern and Western camps "to destroy Islamic movements and fight Islam everywhere." America, as the richer and stronger partner in the conspiracy, had ignited a new "crusading spirit" to obliterate Islamic power and unity and, with Israel at its side, would do its best not to "allow us to establish an Islamic government anew and resume a righteous Islamic life unless we struggle harder."

Fifty years later, one of Qutb's most dedicated disciples and descendants, Ayman Zawahiri, repeated these same charges against America, almost verbatim, in that recruitment videotape I mention at the beginning of this book.

Hence did America become the "Great Satan." Until the morning of November 4, 1979, however, the phrase had remained mostly rhetorical—a denunciation rather than a battle cry. That morning a crowd of Iranian students milled about outside the gates of the American embassy in downtown Tehran. Anger mounted in the chilly air. Very soon the crowd of chanting students, enraged over U.S. support for the dictatorial shah, turned into a surging mob. Panicked, the American sentries retreated into the embassy grounds and then into the embassy building itself when the steel gates gave way. Armed with sticks and stones and a few pistols, students swarmed into the compound. Within a few hours they had created history by capturing and holding hostage an entire diplomatic corps, consisting of seventy-two people.

Never had a Western power suffered similar humiliation. For 444 days the embassy personnel were imprisoned in their own quarters. The American public was riveted by the story, in part because of nightly updates about it on a new show called *Nightline*, which was dedicated to following the crisis in lurid detail. Jimmy Carter was turned out of office because of the debacle. Ronald Reagan swept into the White House, promising Americans that he would restore their power and prestige abroad.

Reagan was popular with Americans, but nowhere near as popular as the Ayatollah Khomeini proved to be among Muslims worldwide. Few in the West believed that this puritanical old man in black robes who had spent most of his life in exile would last long. They were wrong. Khomeini had tapped directly into decades-worth of growing fear and rage against America's foreign policies and cultural invasion, and he knew precisely how to meld cultural, political, and religious resentment. "The agents of imperialism," he told his followers, "are busy in every corner of the Muslim world drawing our youth away from us with their evil propaganda. They are converting them into Jews and Christians; they are corrupting them, making them irreligious." Khomeini offered images of Muslim children being turning into gamblers, prostitutes, addicts, and alcoholics because of sectarian strife and moral corruption.

Khomeini's words and deeds encouraged militants in Lebanon, Kuwait, and Bahrain to kill and kidnap Americans and Westerners. Civil war–torn Lebanon, in particular, proved susceptible to Khomeini's anti-American foreign policy rhetoric. Following the "line of the Imam," Lebanese Shiite commandos perfected suicide or "martyrdom" bombings against U.S. and Israeli targets. Sheikh Mohammed Yazbek, a radical Lebanese Shiite leader, gave a speech praising the 1983 bombing of the marine compound as a "noble action because it shook America's throne. Let America, Israel, and the world know that we have a lust for martyrdom and our motto"—'Death to America'—"is being translated into reality."

Khomeini expanded the Islamists' war against the United States, more than a decade before Al Qaeda began carrying out its suicide attacks against "the head of the snake." The difference between Khomeini's Shiite commandos and Al Qaeda's Sunni jihadists is that bin Laden brought war to

American shores. In effect, Khomeini had blurred the distinction between the "far enemy" and the "near enemy." The "far enemy" was the United States—the Great Satan—and the "near enemy" was its proxy, the Shah of Iran. When his followers overwhelmed the American embassy—sovereign American territory—they were attacking the United States while keeping the struggle local. The tension upon which *Nightline* thrived was whether the students would cross the line and execute the embassy personnel, which would have been an overt act of war.

For the time being the jihad was still a war of rhetoric. Kamal Habib, though an Egyptian Sunni, listened attentively to the Shiite imam from Iran. Many had talked about restoring the sacred mission of Allah, of reanimating the ummah; this old man and his student followers had actually done it. "Ayatollah Khomeini taught us invaluable lessons about dignity, sacrifice, and defiance," Kamal told me on one of our evening strolls through Cairo. "He was not intimidated by the military might of the United States and called its bluff when it threatened to intervene and release its spies and diplomats from captivity. Ayatollah Khomeini stood tall in the face of tyranny at home and domination abroad. If he could make it happen, so could we." Khomeini had demonstrated that "America was a paper tiger," Kamal said, smiling. "When challenged decisively by an authentic leader like the Ayatollah, America failed the test of willpower."

If Khomeini was the first to win worldwide recognition for denouncing what he saw as America's colonialism, Kamal was one of the first leaders to explain to his followers what American foreign policy meant to jihadism. In 1980—a year after the Iranian revolution and a year before Sadat's assassination—he wrote a manifesto called "America, Egypt, and the Islamist Movement," which he secretly distributed it to members of al-Jihad. Echoing Qutb and Khomeini, Kamal ranked the United States as enemy number one, and listed three reasons why. The first was the unholy alliance between America and Arab rulers, which led to Islamic nations' loss of "political, economic, and military independence." Second was America's relationship with Israel, which came at the expense of Muslims, particularly the Palestinians. Third was American global dominance—economic, military, and cultural—which threatened the future of the jihadist movement itself. American hostility to Islam and Muslims, he wrote, was based on religion and ideology. "Crusading hatred drives all American policies toward the Islamist movement." His choice of *crusade*, of course, was not coincidental. He told his cohorts that all Americans, not just politicians, are anti-Muslim and support their government's war against Islam; jihadists must ready themselves for the coming battle because the United States viewed their growing strength as a threat to its presence in the region.

Nonetheless—and it is critical to note this yet again—while Kamal called for cleansing the Muslim world of corrupting Western influences, he stopped short of advocating a direct clash with the United States. Instead, Kamal encouraged attacks on local, secular Arab and Muslim rulers—"traitors" who were doing America's bidding. Kamal was picking up on the rhetorical war Khomeini had started by saying, among other things, that the most effective means to defeating America was "to shed more blood, offer more martyrs and to carry the banner of Islam in order to restore the caliphate or face martyrdom." However, this was still war by proxy. If apostate Muslim rulers could be replaced with authentic Islamic governments, there would be no need to embark on a battle with America itself. As one of Kamal's fellow Islamists, Mamdouh Ismail pragmatically put it to me: "Why should Islamists take high risks by militarily attacking the United States, the unrivaled superpower, if they can achieve their goal by overthrowing ruling Muslim apostates?"

When I asked Kamal about his 1980 manifesto, his views on America had evolved; he no longer lumped all Americans together. "America is definitely a complex and diverse society. There are multiple currents and viewpoints. There are many many progressive Americans with whom we can build cross-cultural alliances. Neither America nor the ummah is a monolith," Kamal added. I was surprised to hear Kamal talk about the existence of common economic interests, common cultural ties, and common geographic links between the Western world and the Muslim region. "We have always had economic, political, and diplomatic relations between us," he intoned. "The historical pattern has been cooperation, not confrontation. Conflict is the exception to the rule."

As Kamal sang the praise of intercultural dialogue and bridge building, he sounded more like Mohammed Khatemi, the recent liberal president of Iran, than either Khomeini or Qutb. The clash of civilizations had been replaced with dialogue. "We are ready to meet America halfway as long as U.S. politicians respect our Islamic identity; stop propping up corrupt, unjust Muslim rulers; and cease their preponderant support for the Zionists," Kamal said, listing the three most important sore points in U.S.-Muslim relations. In this, Kamal had moved away from the jihadist fringe and joined the Arab and Muslim mainstream.

Kamal's rhetoric had softened. That of other jihadists had hardened. Like Kamal, Zawahiri was inspired by the words of Qutb and Khomeini. In his memoir serialized in *Asharq Al-Awsat* after September 11, he echoed them both in his justification for attacking the United States: "The Western forces that are hostile to Islam have clearly identified their enemy. They refer to it as Islamic fundamentalism...It is clear that the Jewish-crusader alliance, led by the United States, will not allow any Muslim force to reach power in any of the Islamic countries." Zawahiri was still pushing the proxy argument of near versus far enemy—that the masters in Washington were using cronies in the Muslim world to protect their interests and fight the battle against the Islamic vanguard on their behalf.

What should be done in the face of this alliance led by the United States against the Islamic movement? According to Zawahiri, "The struggle for the establishment of the Islamic state cannot be launched as a regional struggle...Therefore, to adjust to this reality we must prepare ourselves for a battle that is not confined to a single region, one that includes the apostate domestic enemy and the Jewish-crusader external enemy." In other words, the battle must now be moved to

American soil, "to burn the hands of those who ignite fire in our countries." This meant that the battle would turn into "clear-cut jihad against the infidels."

Zawahiri's discovery of *al-Adou al-Baeed* came late. From the late 1960s until the late 1990s, Kamal and Zawahiri's generation—they are roughly the same age, of course; Kamal and Zawahiri had been among the three hundred activists rounded up after Sadat's assassination in 1981—was preoccupied with the fight against *al-Adou al-Qareeb*, Muslim rulers. In the early 1990s Zawahiri traveled to California's Silicon Valley to raise money from émigré Muslims for his local jihad against the Egyptian regime. America as a target did not figure on Zawahiri's radar screen until he joined Al Qaeda in 1998.

Today, jihadists still look upon America as a threat and the embodiment of evil, but there is no resolution of the "near" and "far" division. In my interviews with radical Islamists I have yet to meet a single one—or read an account by one—who has anything positive to say about America. The closest I came was Kamal's admission, echoing something Qutb had written at one point, that America was a "complex and diverse society," and his statement that even the Great Satan was not, after all, a monolith. Secular nationalists and enlightened Islamists tend to separate American foreign policy from American society and culture. Jihadists, on the whole, do not make that distinction.

This does not mean that actions stemming from a unilateral viewpoint have themselves been unilateral. As much as Qutb, Khomeini, bin Laden, and Zawahiri have reviled the spiritual and political character of the United States, other radicals have found reason to maintain strategic collaboration with the Americans. Throughout the 1970s, '80s, and early '90s an array of Islamic fundamentalists (including the Muslim Brotherhood, as I've suggested) found it beneficial to cooperate with America against atheism and secularism. Wahhabis/Salafis have always deeply mistrusted American foreign policy but were much more intrinsically hostile to communism and secular Arab nationalism (they are, moreover, philosophically disposed toward capitalism, especially private property, free trade, and wealth accummulation). Islamic activists were also preoccupied with bloody power struggles against secular military dictators like Nasser in Egypt, the Baathists in Syria and Iraq, and socialists in Algeria; they feared and loathed those apostate tyrants who were allies of the Soviet Union, who brutally clamped down on Islamic fundamentalists. Forced to choose between the pro-Western camp or the pro-communist bloc, Islamists chose the former—the lesser of two evils.

For example, the Russian invasion of Afghanistan in 1979—the same year as Khomeini's revolution in Iran—reminded Americans and Wahhabis/Salafis of their common interests. For the next ten years Americans and militant Islamists found themselves in the same trenches. To American strategists, the struggle against communism dwarfed the recent feud with the Ayatollah Khomeini. American policy under Reagan remained wedded to the rollback of "the evil empire." As Russian troops marched into Afghanistan, American leaders, who had been caught napping, moved swiftly to mobilize Islamic resistance and to tap into the anti-communist feelings of the now-dominant fundamentalist clergy. Containing Soviet communism, confided Zbigniew Brzezinski, national security advisor for President Jimmy Carter, dictated an avoidance of anything that could split Islamic opposition to the Russians, especially an American-Iranian confrontation. "It now seemed to me more important to forge an anti-Soviet Islamic coalition."

An American-Islamic coalition wasn't particularly new. From the 1950s to the late 1970s, Islamists and American foreign policymakers were allies of convenience in a broad campaign against both Soviet influence and radical Arab nationalism. Whenever the Cold War heated up, American leaders—Republican and Democratic alike—looked at political Islam as an effective defense against the rising local forces of revolutionary secular nationalism and socialism. They looked to traditional Islam to build, as President Dwight Eisenhower said in 1958, an alliance of Islamic states to counterbalance "godless communism" and radical nationalism led by Egyptian President Nasser. In the struggle for world supremacy, the United States used Islamism to neutralize Marxism.

Thus, despite the humiliation in Tehran, the United States spent a decade financing and arming the Afghan mujahedeen, encouraging the recruitment and flow of foreign fighters, including jihadists, into that war-torn country. The Carter and Reagan administrations hoped to harness the religious and ideological fervor of Islamic fundamentalists, paying little attention to the potential militarization of Muslim politics or the rise of a new generation of warriors who could wreck international peace. There was no recognition that the Afghan jihad might spill over, destabilizing Muslim countries as near as Pakistan and as far as Indonesia. Expediency, in short, proved valuable to both sides of the spiritual divide in the fight against Soviet Russia. American strategists saw Afghanistan as a sort of Soviet Vietnam, a long, hopeless, expensive war that would drain Russian will and encourage chaos (which, of course, is exactly what happened). Jihadists found that Afghanistan provided them a safe haven to regroup and gain field experience, recruit new foot soldiers, and build networks between various Muslim countries.

For some years now, Afghan veterans like Zawahiri and bin Laden have tried to rewrite history, going to great lengths to deny having ever had received financial and logistical assistance from the United States and its pro-Muslim ruling partners. They portray themselves as having always been implacable enemies of the Great Satan. For example, in his aforementioned memoir, Zawahiri maintained that all along he and his cohorts in Afghanistan had been training for the coming battle against the United States. "The jihad was a training course of the utmost importance to prepare Muslim mujahedeen to wage their awaited battle against the superpower that now has sole dominance over the globe, namely, the United States." Al Qaeda's official historian labors hard to convince Muslims that jihadists in Afghanistan never dealt with America and its allies, but relied strictly on their own and other Muslims' resources. Though he acknowledges that the United States backed Pakistan and the indigenous Afghan mujahedeen factions with money and equipment, he insists that the mujahedeen had not financed themselves from popular contributions but carried Muslim donations to the Afghan fighters. "Osama bin Laden informed me," Zawahiri writes, that "the size of the popular Arab support for the Afghan mujahedeen amounted, according to his sources, to $200 million in the form of military aid alone in ten years. Imagine how much aid was sent by popular Arab organizations in the non-military fields... Through this unofficial popular support, the Arab mujahedeen established training centers and centers for the call to the faith."

Zawahiri's account flies in the face of evidence. Between 1979 and 1989, estimates are that the Americans and Saudis provided up to ten billion dollars-worth of secret assistance to rebel groups in Afghanistan; the Saudis also partially financed Sheikh Azzam's Services Bureau. Bin Laden was in fact the middleman between the Saudi government and Azzam's guesthouse, and he became the financier of the latter's activities, thanks mainly to Saudi funds and donations flowing through charities or other nongovernmental organizations (NGOs). As the first prominent Saudi to go to Afghanistan, and due to his family's wealth, bin Laden became Saudi Arabia's point man during the Afghan jihad. He only severed his links with the Saudi royal family after the 1990 deployment of American forces in the kingdom. When an Al Jazeera interviewer pressed bin Laden about his former connections with the CIA in the struggle against the Soviets in Afghanistan, he retorted by saying that the Americans were lying when they claimed they had helped. "We challenge them to present a single shred of evidence to prove it. In fact, they [Americans] were a burden on us and the mujahedeen in Afghanistan, and there was no agreement between us.... Unintended confluence of interests does not mean there is any kind of link or tacit agreement."

Zawahiri and bin Laden's belated effort to deny the official Saudi, Pakistani, and American connection is a direct consequence of their decision to go global with jihad and target the United States and its local allies, including Saudi Arabia. Zawahiri is correct to reject the claim that the Arab Afghans were funded (not even "one penny") or trained by the United States. Azzam and bin Laden had access to a broad network of official and semiofficial Gulf funds, as well as to donations by Arab and Muslim NGOs. They did not need American government money to wage jihad, though their network raised funds in the United States. However, their journey was facilitated by America and its Muslim allies; they were part of a diverse and disorganized Islamic army haphazardly mobilized and assembled to roll back the Russian advance. For a decade, jihadists like Azzam, bin Laden, and Zawahiri willingly accepted the Cold War game and played by its rules.

Zawahiri greatly simplifies his and bin Laden's jihadist journeys, which have had many twists and turns. All along, he insists, they have followed a master plan: to join the Afghan jihad caravan, defeat the Russians, and then turn their guns against the only surviving superpower. All this was conceived far in advance, the strategic vision always focused on cleansing Muslim lands of corrupt Western influences and "renegade rulers who allied themselves with the foreign enemies of Islam."

A closer look reveals a more modest, pedestrian track, one lacking any international scope. For most of his life Zawahiri had stressed fighting Muslim rulers and avoiding the distractions of external adventures. As late as 1995, he dismissed Muslim critics who urged him to stop bloodletting at home and, instead, lend a helping hand to the Palestinians by attacking Israel. In a rebuttal entitled "The Road to Jerusalem Goes Through Cairo," published in the *Al-Mujahidun* newsletter, he stated that "Jerusalem will not be liberated unless the battle for Egypt and Algeria is won and unless Egypt is liberated." Even in his own highly revisionist memoir Zawahiri writes that his initial intention in Afghanistan was to "establish a secure base from which to continue to wage jihad in Egypt"—not against the United States, not even against Israel. Neither bin Laden nor Abdullah Azzam viewed the Afghan jihad as a stepping-stone to attacking the United States.

The decision to target America occurred sometime around 1993 to 1994. It was then that they shifted gears from the near to the far enemy. The road to Jerusalem no longer passed directly through Cairo, Algiers, Amman, or Riyadh. Instead it went along an expressway that included stops in Washington, New York, Madrid, London, and other Western capitals.

Why did international jihadists like Zawahiri and bin Laden take the war to America, having already been beaten by the governments of Egypt, Algeria, and Saudi Arabia? Three factors propelled them on this collision course. First, the Afghan war and the humiliating withdrawal of Russian troops in 1989 gave birth to a new, hardened, and professional army of fighters who were habituated, if not addicted, to the business of jihad. The Afghan veterans I spoke with were militarized by their war experience there, especially by their contacts with extremists like the Egyptian Tanzim al-Jihad and al-Jama'a al-Islamiya, the Algerian Islamists, the Pakistani Jamaat-i-Islami, the Kashmiri Harakat ul-Ansar, and others. They had been transformed by the jihad years. They said they were ready to take on the internal and external enemies of Islam. By the early 1990s their ranks had become swollen with thousands of highly motivated young recruits. As a senior Yemeni jihadist confided, "We were on a roll, while powerful Arab rulers were fighting for their political survival."

Bin Laden realized that poorly equipped but dedicated men, armed with faith, could take on better-equipped adversaries and win. In that now infamous three-hour recruitment videotape circulated before September 11, bin Laden calls on Muslims to join jihad against "crusaders and Jews." As the video cuts away to a scene of fleeing Russian soldiers and a Russian personnel carrier destroyed by Chechen mujahedeen, he declares that the Afghan war provides a model for terrorizing the other surviving superpower.

Though Zawahiri held his focus on Egypt and the Middle East until at least 1995, bin Laden's gaze had been steadily shifting westward—again, due in large part to Saddam Hussein's invasion of Kuwait. Immediately following the invasion bin Laden had gone to the Saudi royal family with a personal proposal. He would mount a force of 100,000 trained mujahedeen to enter Kuwait and expel the Iraqis. The Saudis listened but, pressured by the Americans and hardly keen about having a large army of volatile jihadists on their doorstep, declined bin Laden's offer. Worse, the royal family agreed to the permanent stationing of American troops on holy ground.

For bin Laden, that was the point of no return, and the second factor in his decision to take the jihad to America. Overnight, the United States went to the top of his list of enemies. He called on young Saudis to initiate a guerrilla war. However, he steered clear of advocating the overthrow of the Saudi royal family, cautioning his supporters to avoid the trap of "an internal war." Local regimes, including the house of Saud, had become insignificant tools, mere agents of the American-Israeli alliance that kept the ummah "divided." Saudi Arabia was therefore to be considered an occupied country, its regime incapable of forcing the Americans out. Once the United States had been expelled from the Land of the Two Holy Sanctuaries, its vassals would fall like rotten fruit.

"The wounds of the Muslims are deep everywhere," bin Laden intones in that recruitment video. The tape opens with images of American troops in Saudi Arabia, intercut with clips of American presidents fraternizing with Saudi leaders. Bin Laden laments that Saudi rulers have allowed American troops, including Jewish and Christian male *and* *female* soldiers, to set up camp on the soil where the Prophet Mohammed was born and the Qur'an descended: "This land is exceptional because it is the most beloved by Allah. How could it be that the Americans are permitted to wander freely on the Prophet's land? Have Muslim peoples lost their faith? Have they forsaken the Prophet's religion? Forgive me, Allah, I wash my hands of these [Muslim] rulers!" Horrifyingly bloody images of Palestinian, Iraqi, and other Muslim children pan across the screen, while bin Laden angrily demands, "Where is the Muslim ummah and its one billion believers? The ummah sees and hears that the Qur'an is being defamed, burned, and used by the Jews as disposable tissues, yet it stands idly by."

On August 7, 1998 Al Qaeda struck against America in simultaneous suicide attacks on the embassies in Tanzania and Kenya. Two-hundred and fifty-seven people were killed and five thousand injured; twelve Americans were among the fatalities. The bombings demolished the embassies. President Clinton was forced to strike back and try to "wipe out Al Qaeda leadership." On August 20, seventy-five missiles swept across over the Pakistani night sky at about four hundred miles an hour, hitting Al Qaeda camps in Afghanistan. Other Tomahawks were also fired against the al-Shifa chemical plant in the Sudan. Neither bin Laden nor his senior aides were killed. Far from deterring bin Laden, Zawahiri, and their fellow jihadists, Washington's response emboldened them. At last the war with the United States had arrived. If and when the Americans invaded Afghanistan, as they expected they would, the jihadists would turn the country into a graveyard, just as they had for the Russians. Shortly after the U.S. missile strikes, Zawahiri got on the phone to the Pakistani journalist Rahimullah Yusufzai, who had interviewed Al Qaeda and Taliban leaders many times, to declare that he and bin Laden were safe. "The war has only just begun," he stated.

Thirdly, the decision to internationalize jihad and attack the United States occurred after pro-Western rulers had militarily and bloodily suppressed Islamist uprisings in Egypt, Algeria, and Saudi Arabia during the second half of the 1990s. Jihadists blamed Western powers, particularly the United States, for tilting the balance of power in favor of local rulers; they argued that the West, led by America, was intrinsically hostile to the establishment of authentic Islamic states and would do everything in its power to nip the movement in the bud. "America played a big role behind the scenes in the fight against the Islamist movement," Mamdouh, the Islamist attorney, told me vehemently. "We are part of a world order managed and controlled by imperial America. Nothing happens without U.S. knowledge and intervention."

As I indicated at the beginning of this book, by the time I arrived in Cairo in 1999 the local jihadists had called it quits. During the two years I was there, however, a major struggle broke out between local and international jihadists. For the first time the two camps bickered openly over the leadership and future direction of the movement. Kamal and his friends spearheaded the first camp and were trying almost desperately to end the state of war between local jihadists and their governments; they wanted, as they told me, to formalize the unilateral ceasefire declared by their associates in 1997. By contrast, bin Laden, Zawahiri, and company were pursuing bigger ambitions—waking the Muslim community from its slumber. Attacking the United States would reinvigorate the jihadist movement and build its credibility in the eyes of Muslims worldwide; it could even trigger a clash of cultures. In a secret 1998 letter to another militant—recovered in 2001 from captured Al Qaeda computers in Kabul—Zawahiri points out that Al Qaeda had escalated the fight against "the biggest of the criminals, 'the Americans' to drag them for an open battle with the nation's masses..."

The near enemy and the far enemy had become as one, a great world order that needed to be destroyed. As the global enforcer, the Great Satan stood for all the evils rising up against the ummah. "We must completely topple the United States and we hope to be the ones who can topple its entire system," bin Laden told his followers.

V

Under Middle Eastern Eyes

"HOW COULD NINETEEN young Arab men fool the CIA?" That was the question repeated over and over when I arrived in Beirut two weeks after September 11 to attend a meeting organized by the Centre for Arab Unity Studies, a pan-nationalist think tank. A broad spectrum of scholars, political leaders, and activists had come together to assess the likely repercussions in the Muslim world of the recent attacks.

This was not to prove a simple matter. No sooner had I landed at Beirut International Airport than I entered into a whirlwind of skepticism, distrust, and conspiracy. I also wanted to talk to Islamists and ordinary people and get a sense of how they viewed what had happened. From airport workers to taxi drivers to bank managers to university students, no one believed that Arabs could have been responsible for September 11. Still in awe of American power and the CIA legend, ordinary Arabs and Muslims simply could not imagine that Al Qaeda could have been responsible for the attacks. Years would pass before some would admit to recognizing Al Qaeda's fingerprints on the September 11 crime scene—even after bin Laden and Zawahiri had publicly boasted about their feat.

Most of the people I spoke to had other theories. The top choice was circulating on various Western Web sites, pointing the finger at Mossad, the Israeli intelligence service. Another speculated about rogue elements within the U.S. government itself. Secularists and nationalists were as skeptical as Islamic activists about the accusations being leveled by the Bush administration against Al Qaeda. A Christian manager of a European bank in the Ashrafiya district of Beirut, told me that only "international Zionism possessed the means and the will to undertake this hideous act." Then he paused knowingly. "Everyone knows this." A shopkeeper with Islamic sympathies in the coastal town of Sidon dismissed the case against Al Qaeda out of hand. "Who benefits from these attacks?" he asked rhetorically, "the Arabs or the Jews? Who has vested interests in causing a confrontation between Muslims and Americans? Those who benefit the most—the Israelis and their allies—plotted and carried out the bombings. How else do you explain that Jews were not among the thousands of casualties? Plus, our religion forbids us to harm civilians."

Everyone—whether bus driver or college teacher—saw a nefarious conspiracy. Most striking to me was an encounter

I had almost a year after September 11 at a reception at the American University of Beirut (AUB) which is a liberal enclave in the Arab world. A young Ivy League–educated philosophy professor argued passionately that four Arab pilots could not possibly have carried out the attacks on the United States, because it would have taken a team of seasoned engineers to score direct hits against the World Trade Center and the Pentagon. He also looked to conspiracy. A senior administrator at AUB later told me that the professor's comments reflected the predicament of young Arabs who feel politically impotent and powerless. However irrational, the attitudes of Arabs and Muslims were shared by a great many Westerners and Asians. Most often they linked the September 11 attacks to American foreign policies toward the Palestinians and Iraqis. In short, they believed, America had reaped what it sowed.

Few of the participants at the Centre for Arab Unity conference were convinced of Muslim culpability in the September 11 attacks. The consensus was, "let us wait and see the existing evidence" before rushing to indict Al Qaeda. Compounding the skepticism were those deeply held suspicions about U.S. foreign policy objectives. Muslim opinion makers clearly believed that the United States was using September 11 as a justification to destroy bin Laden's Al Qaeda organization and to topple the Taliban regime in Afghanistan. Almost unanimously, they believed that the Bush administration had an overarching strategy for winning control of the oil and gas resources in Central Asia, encroaching on

Chinese and Russian spheres of influence, destroying the Iraqi regime, and consolidating America's grip on the oil-producing Persian Gulf regimes.

Many Muslims suspected the Bush administration of exploiting September 11 to settle old scores and to reassert American hegemony in the world—all this despite the fact, unacknowledged by the Beirut participants, that the terrorist attacks had forced the new president to shift his focus from a domestic agenda to world affairs. Regardless of President Bush's efforts to allay the fears of Muslims by stressing that the United States would wage a relentless war only on terrorists and those that "harbored" them, that message seemed to have fallen on deaf ears. Liberals, leftists, and Arab nationalists sounded as mistrustful of American aims as the representative from radical Hizbollah. Most of the participants, who represented the pulse of mainstream Muslim opinion, strongly cautioned against joining the anti-terror coalition. They also warned that they would oppose any sustained military assault on a Muslim country, including Afghanistan.

I tried to remind my colleagues that America was no longer the same country it had been before September 11. It was now an injured superpower with injured pride. Bush would pursue aggressive, militaristic policies to send a message of resolve to the world as well as to reassure an anxious citizenry. I tried to convey to my colleagues that Muslims and Americans were at a crossroads—the dawning of a new and dangerous era. Arabs and Muslims should unequivocally condemn Al Qaeda and join the efforts by the international community to complete its encirclement; the unprovoked attack on American civilians was a crime against humanity that demanded an appropriate response.

Hizbollah's representative, Nawaf al-Musawi, a rising star within the party who is in charge of foreign relations, was not impressed by my plea. "We condemn the killing of civilians, all civilians—Americans and Muslims alike," he responded heatedly. Broad-shouldered and sturdily built, Nawaf struck an imposing figure. "American blood is not thicker than Arab and Muslim blood," he railed, pounding the table with his fist. "Why were all those voices of humanity and morality silent when Israel bombed and massacred Lebanese and Palestinian civilians? Where was American public opinion and conscience? We feel there are two standards of morality being applied in international affairs—one for the Americans and the West and one for the rest." I looked around the room. The other participants were listening attentively to Nawaf's speech and nodding their heads in agreement.

I sought out Nawaf during lunch. I wanted to hear exactly how Hizbollah viewed the attacks on New York and Washington and their potential reverberations. First, he wanted me to understand clearly that his party had no ties to Al Qaeda whatsoever and possessed no global ambitions. "Our Islamic movement is dedicated to expelling the Zionist occupiers from the remaining farms in southern Lebanon and to keeping Israel at bay. We are a local resistance party and conventional political organization with a very specific program," he told me as we ate lunch overlooking the Mediterranean. "I do not know if Al Qaeda carried out the attacks on New York and Washington. I have no faith in what U.S. rulers say. But we, as a party, do not condone or encourage the killing of civilians—any civilians. Americans are not unique. They should not be treated differently from other victims, including hundreds of thousands of Muslims who perished in Lebanon, Palestine, Iraq, Iran, and Kashmir. History neither begins nor ends on September 11. Americans have no monopoly on suffering and pain." He added that he hoped America would use the attacks "as a catalyst to build bridges to Arabs and Muslims and begin the healing process," but that he doubted this would be the case. "My instincts tell me that Bush junior will exploit the loss of American life for domestic political advantage and will go militaristic."

Nawaf was not the only Hizbollah official to argue along these lines. All the Hizbollah members I met in Beirut categorically denied having had relations with any terrorist organization, including Al Qaeda; they were laboring hard to shed their party's violent past and project a new peaceful image. They did not mince words about their loathing of American foreign policy, but emphasized that bin Laden was not one of their own and that his fight was not theirs. Not one uttered a single word of praise for bin Laden. This was perhaps not surprising, given that most members of Hizbollah are Shiite, and bin Laden is a militant Sunni Wahhabi/Salafi who considers Shiites less Muslim than himself. Haytham Mouzahem, a young Shiite journalist who worked for the Arabic newspaper *Al-Mustaqbal*, and who has close contacts with Hizbollah, told me not to be surprised by the party's lack of empathy and support for Al Qaeda. The two militant organizations are more "foes than friends," he said, and possess differing political agendas and priorities. While both are hostile to America, Hizbollah was not as "reckless" or "suicidal" as bin Laden's predominantly Sunni network. "Hizbollah will not shed a drop of blood on behalf of Al Qaeda," Haytham said as we sat nibbling on sweets at Al-Halab, West Beirut's most popular pastry shop, facing the Mediterranean.

Nawaf and Haytham proved correct. The Hizbollah leadership distanced itself from September 11 and went public in its criticism of Al Qaeda. Sayyed Mohammed Hussein Fadlallah, Hizbollah's spiritual founding father, dismissed Al Qaeda's claim that its attacks had been religiously sanctioned. In dozens of interviews, sermons, and lectures immediately after September 11, Fadlallah, one of the most prominent radical Shiite clerics, called Al Qaeda's bombings "suicide" rather than "martyrdom operations," and thus deemed them illegitimate. Even while he vehemently criticized (of course) American foreign policy, Fadlallah flatly opposed killing American citizens, who were not responsible for their country's international affairs, and may even have opposed U.S. policies. "We must not punish individuals who have no relationship with the American administration or even those who have an indirect role."

A distinguished white-bearded cleric in his seventies, Fadlallah's defiance of (and some say incitement against) Americans in Lebanon during the 1980s earned him great respect across the spectrum of Sunni and Shiite public opinion. Although he denies Hizbollah involvement in any terrorist activities, he has justified the earlier targeting of Americans and Westerners in Lebanon because they directly aided the Israeli occupation of the country. In his speeches after September 11 Fadlallah stressed several factors. First, not only should Americans not be equated with their government but that the so-called clash of cultures was a fiction. Fadlallah referred instead to a "struggle against arrogance," a code for American hegemony. Bin Laden, he said, should remember that "civilized Islam" does not condone preemptive strikes against citizens of nonbelligerent nations like the United States, even if its policies harm Muslims. Faulting bin Laden's grasp of political realities in the world, he reprimanded Al Qaeda for failing to consider the costs and benefits of its actions and suggested that in fact Al Qaeda did not care about the damage it had caused the ummah.

Fadlallah rejected the official U.S. comparison between Al Qaeda's attacks on Americans and Palestinian attacks on Israelis in the occupied territories. Palestinians were justified in carrying out "martyrdom operations" against Israeli military and civilian targets, he argued, because Israel occupies Muslim lands and oppresses believers; Palestinians were doing nothing more than defending themselves. In that context he characterized martyrdom as the Palestinians' most effective deterrent. September 11, in contrast, could not be religiously or politically sanctioned because America neither directly occupies Muslim territories nor oppresses Muslims. Fadlallah's critique of Al Qaeda resonates strongly with radical young Muslims today. Bin Laden and Zawahiri are in deep trouble when a revolutionary cleric like Fadlallah unequivocally repudiates their actions and calls on believers to exercise restraint and refrain from attacking American civilians.

Haytham told me to pay more attention to Sunni militants than to Shiites. "The current wave of jihadism is overwhelmingly led by Sunni Wahhabis," he said, "while militant Shiism dominated the Muslim scene throughout the 1980s." Unlike their Shiite counterparts, Lebanese Sunni Islamists immediately voiced support for the September 11 attacks. Fathi Yakan, the leading ideologue of al-Jama'a al-Islamiya, called the September 11 attacks a "heavenly blow" against American symbols of power, telling his followers that the United States is making war against "Islam and Muslims everywhere" under the guise of fighting terrorism.

Hicham Shihab, a former al-Jama'a fighter, advised me to meet with al-Jama'a members and other Sunni militants in Tripoli and Sidon. A taxi driver named Abu Bilal volunteered to drive me down from Awkar, a northern Beirut suburb, to Tripoli, and introduce me to a few of his "mujahedeen friends" who had fought in Afghanistan alongside bin Laden and Zawahiri. A small contingent of these Al Qaeda–trained veterans lived in Tripoli; al-Dinniyah, a mountainous area east of Tripoli; Sidon; and in the Palestinian refugee camp of Ain al-Hilweh, located on the outskirts of Sidon. Abu Bilal promised to drive me there and bring me home safely. Before

we left, he strongly advised me against asking the mujahedeen about their real identities: "You do not want to raise
their suspicions," he said, winking.

Once we were on our way, Abu Bilal volunteered without hesitation that America's "oppressive" policies left it
"with no friends—only enemies" among Arabs. He told me
that he had gone from being a Nasserite to an Islamist in the
1970s. "We felt we needed to go back to our Islamic roots to
be able to stand up to the Zionists and Americans. Nothing," he said, turning to face me, "could stop those young
Arab men who piloted the martyrdom planes," his voice
rising as he uttered the word *martyrdom*. The ride from
Beirut to Tripoli takes about an hour and a half—plenty
of time for Abu Bilal to lecture me about the evils of Western colonialism and America's arrogance and oppression
of Muslims. "I do not understand why Americans are surprised that we finally retaliated against them," he said. "We
could no longer tolerate injustice. Armed resistance is the
only language that the powerful understand."

"Do you really support the targeting of civilians?" I asked.

He hesitated for a moment before answering and stared
at me inquisitively. "It is a difficult question because Islam is
opposed to killing civilians, especially women, children, and
the elderly. You must know that. But ask the Americans why
they starved millions of Iraqi children to death. Palestinian
children are being shot in cold blood by the Jews. Our pa-
tience has been exhausted. Our cries for justice have gone
unanswered." Abu Bilal's anger mounted steadily. We were
driving along the beautiful coast and I began to grow anx-

ious. He kept taking his eyes off the road and his hands off the wheel to make a point. I breathed a sigh of relief when we made it safely to our destination in Tripoli.

We arrived on a typically sunny day around ten in the morning and headed toward Bab al-Tubaneh Street, which is situated in the poorest neighborhood in the city. Many of the dilapidated concrete apartment buildings bore the marks of the fifteen-year civil war, their walls scarred by artillery and mortar shells. Bab al-Tubaneh reflects the cultural and architectural decline of Tripoli. Once one of the richest and most renowned Islamic cities—the "jewel of the Arab East," it was called by some—it was ransacked a thousand years ago by marauding Christian crusaders who made a point of burning its magnificent library, which contained thousands of manuscripts. According to Arab historians cited in *The Crusades Through Arab Eyes* by Lebanese writer Amin Maalouf, it took the crusaders thirteen years to destroy ancient Tripoli—its mosques, palaces, rings of olive trees, and sugarcane fields. By the end of the siege, jihadism, which until then had been little more than a colorful slogan, took on new life and meaning. Refugees and clerics from Tripoli began preaching its necessity. Tripoli never recovered fully from its desecration at the hands of the Christians.

The last fifty years have witnessed another steep decline in the city's fortunes. As happens in so many other developing nations, the Lebanese government pampered the capital, Beirut, at the expense of the rest of the country. It is no wonder that extremism flourished in poor neighborhoods like Bab al-Tubaneh, where unemployment exceeds 30 percent and social services are almost nonexistent. Despite the poverty and decayed infrastructure, however, traces of Tripoli's former grandeur persist. Beneath the dirt and behind the rubble, you can find an eclectic mix of Roman, Islamic, and Italianate styles. Almost every street harbors some architectural gem; garbage infested alleyways open onto beautifully decorated Ottoman-era mosques and palaces, now converted into apartments. Despite it all, Tripoli remains the liveliest city on the Lebanese-Syrian coast.

Abu Bilal parked his car next to a vegetable market jammed with dozens of stands. We walked a few blocks and entered a dimly lit building. Three thickly bearded men dressed in long white robes and prayer caps were seated on floor mats.

"Assalamu Aleykum," Abu Bilal and I greeted our hosts.

"Wa Aleykum Assalam," they greeted us back.

This symbolic Islamic welcome broke the ice. Abu Bilal introduced the men to me as Abu Abir, Abu Mohammed, and Abu al-Walid. He presented me as Dr. Fawaz, a neighbor, born in Akkar, who would like to hear their reactions to the September 11 attacks in America.

"Why are you interested in our views?" Abu Mohammed asked me. I replied that I had come to Beirut to attend a meeting organized by the Centre for Arab Unity Studies and that I thought it would be useful to listen to what they had to say. I explained that I had already spoken to members of Hizbollah. Now I wanted to know what they thought.

"Our view is the same as Sheikh Abu Abdullah"—Osama bin Laden—Abu Abir volunteered calmly. "Sheikh Osama said that he had no hand in the explosions in New York and Washington. We believe him. If he had ordered the attacks, he would have taken responsibility. He is no coward like that little [junior] Bush."

Abu al-Walid, a man in his thirties, broke in. He wondered why we were making a big deal about the Americans. "Millions of Muslims have been killed by the Jews and their American masters," he said. "Why was the world not outraged about the Muslim victims? America got what it deserved. God punished it for its crimes against humanity, particularly the ummah."

I asked Abu Abir who he thought had carried out the attacks.

"Who else?" he retorted instantly. "The Jews have been desperately trying to destroy the Islamist movement, especially the Al Qaeda organization, which represents the vanguard of the ummah and a real threat to its enemies. They want to pit America against us—mujahedeen—and dissipate our strength. We are ready to fight and gain martyrdom. Like Communist Russia before it, America will be defeated if it dares to invade Afghanistan." His voice was thick with rage.

"But are you not anxious that you will find yourselves facing the armed might of the unrivaled superpower?"

"Absolutely not!" Abu Mohammed replied. "We are armed with belief. Sheikh Osama knows what he is doing. He was not born today. If and when Afghanistan is attacked, thousands of mujahedeen will join the battle. Bush must know that the ummah will not stand idle when one of its members is attacked. In Lebanon, we have hundreds of brothers who could be mobilized overnight to carry out martyrdom operations in defense of the ummah."

There may or may not be hundreds—or even dozens— of Afghan war veterans in Lebanon, but Abu Mohammed's threat was not to be dismissed lightly. On New Year's Eve 2000, a group of several hundred militants launched an attack on the Lebanese army in al-Dinniyah, east of Tripoli. It took the army more than a week to defeat the insurgents, many of whom had been trained at Al Qaeda camps in Afghanistan, and to kill the Lebanese-born leader of the uprising, Bassam Ahmad Kanj (alias Abu Aisha), an acquaintance of bin Laden. I happened to have been in Lebanon at the time. The whole nation, including mainstream Islamists, was shocked by the militants' show of force, fervor, and fanaticism; they seized control of dozens of villages in the al-Dinniyah district and massacred inhabitants at will. Although the uprising failed, and most of the militants were killed, a band of survivors fled by boat and took shelter in the Palestinian camp of Ain al-Hilweh.

The day after meeting the three mujahedeen I asked Abu Bilal to drive me to Sidon and Ain al-Hilweh. I hoped to see for myself if Al Qaeda had established a foothold in the largest Palestinian camp in Lebanon. More than seventy-thousand refugees live there on a three-quarter-square-mile piece of land with no sanitation or clean water. The refugees mainly depend on the United Nations Relief and Works Agency (UNRWA) for housing, health care, and schools. They have no regular sources of income. Ain al-Hilweh is a testament to the plight of the four million Palestinian refugees dispersed throughout the world, most of them in Arab countries.

Abu Bilal said he had no contacts in Ain al-Hilweh. He advised me not to go there without first finding a person who could call on my behalf and introduce me to militants. I called Hicham and asked him for help. He suggested that I meet with Sheikh Jamal Khattab, an adherent of militant Wahhabism/Salafism of the bin Laden strain, and the spiritual guide to three small pro–Al Qaeda affiliates—al-Haraka al-Islamiya al-Mujahida, or the Islamic Struggle Movement; Osbat al-Ansar, or League of the Partisans; and Osbat al-Nour, or League of Light. Hicham knew him well and told me that Khattab used al-Nour Mosque as his headquarters. The mosque was located in a street known as Hey al-Mawt, or Neighborhood of Death, because it had been the place of so many killings among rival factions.

My father, retired and in his sixties, knows Sidon well. He insisted on accompanying me as I headed toward Sidon and the Ain al-Hilweh camp the following morning. It took almost an hour to reach Sidon, one of the famous names in ancient history ("Saidoon" is the Phoenician name, "Saida" the Arabic). Sidon has been known since Persian times as a city of gardens and orchards, and it is still surrounded by citrus and banana plantations. Today it has the feel of a small town. Like Tripoli, ancient Sidon resisted the crusaders who then took their revenge by ravaging its orchards and pillaging nearby villages. New depredations arrived in the nineteenth century, when European treasure hunters and amateur archaeologists made off with many of its most beautiful and important ancient art objects, some of which can now be seen in foreign museums. In this century, too, ancient objects from Sidon have turned up on the world's antiquities markets. Other traces of its history lie buried beneath the concrete of modern construction. The city's former glory remains barely visible in the beautiful old soukh, the landmark crusader castle, and the modern port installations that grace its entrance.

My father took me directly to the soukh to taste the *serinouras*, the city's famous sweets, and to converse with the locals. It was there that I met Abu Sa'ad, an Islamist who voiced his skepticism about Al Qaeda's responsibility for September 11. I could not find anyone who believed that nineteen Arab hijackers could have carried out the spectacular, simultaneous attacks on the United States. It didn't make sense, I was told, because "we are talking about America—the superpower—not a banana republic."

"No nineteen young Arabs, or even nine hundred, are capable of infiltrating the American security system," one middle-aged customer barked at me. "Who does this Bush think we are? More than once our fingers have been burned by the American fire. Tell Bush that believers will not be deceived twice." He then left the shop without offering the standard good-bye salutation.

We headed toward Ain al-Hilweh, whose four entrances are controlled by the Lebanese army. It looks and feels like a sprawling prison. Internal security is maintained by rival Palestinian groups, ranging from the ultraleft to the ultra-right (and pro–Al Qaeda). After passing through one Lebanese army checkpoint we were stopped at another—five hundred feet away from the first—manned by fighters from Fatah, the military arm of the Palestine Liberation Organization (PLO). I asked for directions to Hey as-Safsaf, the official designation for Hey al-Mawt, and I was pleasantly surprised that the Fatah fighters were neither alarmed nor suspicious. Western journalists must be frequent visitors to Hey al-Mawt, given its concentration of militant Islamists and jihadists.

I have visited Palestinian refugee camps in Lebanon, Jordan, Syria, and Palestine but I was unprepared for Ain al-Hilweh. A few narrow streets run through one of the most densely populated areas in the world. Cars honk their way through pedestrian crowds. Large families of ten or twelve people live in single rooms. As soon as we reached al-Nour Mosque, the crowds thinned. Palestinians avoid it because skirmishes between Osbat al-Ansar and rival factions often flare up without warning. The shacks lining the alley are riddled with bullet holes. There are no sidewalks. The only structure, looming like a fortress over the adjacent hovels, is a two-story building that houses al-Nour Mosque, an elementary school, a small library, and other facilities for teaching the Qur'an. Sheikh Khattab runs the whole venture and is widely believed to be the coordinator of fundamentalist factions. On that day Sheikh Khattab was nowhere to be seen. His students were not forthcoming as to his whereabouts; instead they suggested that I talk with some of their brothers at a nearby restaurant.

Initially, Abu Mohammed, the thickly bearded owner in his late thirties, was very suspicious. He interrogated me about my "mission" and whom I worked for. I told him I was a researcher and I wanted to understand how the brothers viewed the September 11 attacks.

"You mean the blessed attacks," he responded. Abu Mohammed needed no prodding. He boasted about how Sheikh Osama and Dr. Ayman (Zahawiri) had "humbled the head of arrogance" and "bloodied its nose." A few of his customers, bearded and dressed in military uniforms, nodded in agreement.

"Did Al Qaeda carry out the bombings?" I asked, expecting him to deny it like almost everyone else.

"Yes, the mujahedeen did it," Abu Mohammed said without hesitation. "It is a payback for some of America's crimes against the ummah. Look around you. Look at Palestinian misery. Who do you think is responsible for the Palestinian tragedy? Who maintains Israel's military and technological superiority over all Muslims? America."

"But how about the killing of American civilians?" I asked.

Abu Mohammed did not wait for me to complete my question. "How about Palestinian civilians?" he retorted. "How about Iraqi civilians? How about Lebanese civilians [killed during the 1982 Israeli invasion of Lebanon]? In the case of the Jews, America supplies them with the weapons and bullets to kill Muslims. In the case of the Iraqis, America does the killing herself. Al Qaeda is defending the ummah. You should not ask me about American civilians. You should ask Americans how they feel about the killing of Palestinian civilians."

Abu Mohammed and his friends said that they fully supported Al Qaeda and they were prepared to answer bin Laden and Zawahiri's call to wage jihad. When I enquired about their actual numbers and strength, Abu Mohammed was less forthcoming and seemed less confident: "The brothers number in the hundreds." When I pressed him further, he conceded that the overwhelming majority of fighters in Ain al-Hilweh belonged to Fatah and were *kufar,* "infidels," meaning non-Islamists. Fatah members do not support Al Qaeda and have tried to undermine and disarm jihadists like Khattab and his bands. More than once the two rival camps have engaged in deadly skirmishes. I also learned that membership in the three extremist Islamist factions numbered in the dozens, not the hundreds, as Abu Mohammed claimed, and that even they are further splintered into factions.

Later I found out that Abu Mohammed's real name was Abu Mohammed al-Masri, or Mohammed Abdel-Hamid Shanouha, or Abdel-Sattar Jad, an alleged lieutenant in Zawahiri's Tanzim al-Jihad. He was in charge of the Arab Afghan veterans living in Ain Hilweh. On Saturday, March 1, 2003, Abu Mohammed was killed by car bomb that exploded outside his restaurant as he was heading to morning prayers at al-Nour Mosque. Sheikh Khattab accused Israel of the assassination. "The man was targeted personally," he maintained. The booby-trapped, Beirut-registered car was parked near the restaurant on Friday evening, according to Sheikh

Khattab. The driver said he was going to buy cigarettes from a nearby shop but never returned. Several witnesses reported hearing a drone flying throughout the night before the bombing took place.

The one inescapable lesson I drew from my visit to Ain al-Hilweh is that the Palestinian predicament has profoundly radicalized and militarized generations of young Arabs. I am often asked by American journalists why Arabs and Muslims hate American foreign policy. Why the anger and rage? I only wish that every U.S. commentator and reporter would visit Ain al-Hilweh refugee camp to see for themselves how bleak social and political conditions breed extremism and anti-Americanism. I was frankly surprised that Al Qaeda had recruited only a few dozen militants in Ain al-Hilweh; I would have expected to find hundreds, if not thousands, of bin Laden's followers there.

◦═╪═◦

Yet the impressions and reflections I gathered in Beirut, Tripoli, Sidon, and Ain al-Hilweh reveal only a part of the story. Reactions to September 11 differed considerably from one group to another and reflected deep cleavages and differences, clear signs of the internal struggle that is tearing the jihadist family apart—a civil war has hardly been noticed, let alone critically examined, in the United States.

One of the major miscalculations bin Laden and Zawahiri made was in thinking that by attacking the United States they would lure estranged local jihadists like Kamal back into the fold; they believed that they would mobilize believers against pro-Western Muslim rulers and their superpower master. They expected a Muslim response similar to that following the Russian invasion and occupation of Afghanistan. Their goal was to generate a major world crisis, provoking the United States "to come out of its hole," as Seif al-Adl, Al Qaeda's overall military commander, wrote in a 2005 document; American attacks on Muslim countries would reinvigorate and unify a splintered, war-torn jihadist movement and restore its credibility in the eyes of the ummah and of beleaguered people elsewhere. At the heart of their thinking lay that idea advanced by Sayyid Qutb more than half a century before and carried forward by so many others—that only an Islamic "vanguard" could rid Muslim society and politics of ungodliness and restore God's sovereignty.

When the United States invaded Afghanistan, however, Al Qaeda found itself on its own. The river of seasoned jihadists and fresh volunteers it had expected to flow into the Afghan theater turned out to be a mere trickle. Far from rushing to defend their international brethren, most decided not to take sides. America's Afghan war proved to be dramatically different from Russia's. When Russian troops invaded Kabul, the calls for jihad echoed from almost every corner and mosque in Arab and lands. At least fifty thousand faithful flooded into Afghanistan; they had the blessings of the religious and the ruling establishments. A comparatively deafening silence followed the United States' war on the

Taliban and Al Qaeda. Although many Muslims criticized America's impulsiveness and reliance on force, almost none called for jihad against the United States.

No religious authority lent his name and legitimacy to repelling the American troops. In response to an inquiry from the most senior Muslim chaplain in the U.S. army, a group of leading Islamic scholars issued a fatwa on September 27, 2001, directing that American Muslims were obliged to serve in the armed forces of their country, even when the United States was at war with a Muslim nation. Yusuf Qardawi, one of the best known conservative Islamic scholars—someone who has never hesitated to criticize American foreign policies—endorsed the fatwa. The ummah was not on the same wavelength as Al Qaeda.

Private messages and internal correspondence stored on Al Qaeda computers captured in Kabul provide a detailed account of the network's thinking immediately following the September 11 attacks. For example, a draft of an open letter to the American people from October 3 declares that the U.S. government was itself to blame for the September 11 attacks because of its pro-Israeli policies, and it calls on ordinary Americans to be the "ax to break all these chains." Some of the computer messages show how clearly bin Laden miscalculated America's post–September 11 resolve. Shortly before the American bombing campaign, he wrote a soothing message to Mullah Omar, advising the Taliban leader that Washington might shy from military action. Even if America did strike, he added, it would retreat quickly, humiliated like the Soviet Red Army in the 1980s. Either way, he said, America would end up a "third-rate power like Russia." Bin Laden urged a propaganda campaign to convince the American public that intervention would only lead to "further losses of money and lives." He told Mullah Omar this would "cause a rift between the American people and their government." The letter, signed "your brother Osama bin Mohammed bin Laden," was dated October 4, 2001.

Three days later, Bush announced that warplanes had launched strikes against Al Qaeda camps and Taliban positions. After the bombing began, Al Qaeda, evidently hoping for a replay of the anti-Soviet jihad, drafted a budget for raising an army of mujahedeen. Its bookkeepers estimated that it would cost $670,000 to arm and clothe two thousand fighters; each would be given a Kalashnikov rifle, four grenades, two pair of socks, slippers, and a hat.

The messages reveal that panic had already taking root. One militant drafted an e-mail requesting that his cash be moved elsewhere. Bin Laden fretted about his own safety. In a separate letter, he told Mullah Omar that the United States intended to "arrest all those whom America terms terrorists, at the head of which is my weak person." He warned Mullah Omar that he would likely be killed. The letter doesn't appear to have been finished.

In the ensuing weeks, Al Qaeda leaders scrambled to save themselves and their cause. They abandoned grand plans to extend their parallel state in Afghanistan, with its own bureaucracy. Instead, they re-branded Al Qaeda as a champion of Palestinians and Iraqis, people in whose suffering bin Laden hadn't previously shown serious interest.

They also set about salvaging their core activity: suicide terrorism. "Stimulating jihad against the crusader aggression requires a stepping up of incitement to jihad, of training and preparation, of martyrdom and of spending for God's cause," bin Laden wrote in one computer message. Al Qaeda chiefs focused on what they saw as the West's big weakness: its struggling economy. They boasted of having crippled capitalism. In one letter to Mullah Omar, bin Laden reported with glee how "many American and European airlines are on the verge of bankruptcy," and "seven out of every ten Americans suffer psychological problems following the attacks on New York and Washington."

In the *Wall Street Journal*, Andrew Higgins and Alan Cullison reported: "The last retrievable document on the hard drive is a rambling denunciation of the U.S. Apocalyptic in tone and in places barely coherent, it was stored on the computer Nov. 10, three days before Taliban troops fled the Afghan capital and the Northern Alliance moved in, killing some Arabs left in the city. In apparent fury at the failure of Muslims to rally against the West, the tract fumes that 'weapons used against you exist in your home, to entertain you, and to amuse your sons.' America, says the tract, has a 'strong plan—total domination of the peoples' lives.'"

Al Qaeda chiefs were disarmingly candid in a now infamous videotape recorded in Afghanistan immediately after the September 11 attacks and later obtained by the Department of Defense. The video, which was apparently not meant for public consumption, shows bin Laden, along with his senior aides, including Zawahiri, welcoming and chatting with a Saudi militant and sympathizer, Khaled al-Harbi, who had just been smuggled in from Saudi Arabia via Iran. Al-Harbi conveyed messages of support to bin Laden from leading Saudi clerics and inquired about Al Qaeda's thinking and morale. The host and the guest saluted the September 11 suicide hijackers and praised God for bestowing martyrdom on them.

> **bin Laden:** *So these young men, may God accept their actions. Nawaf al-Hazmi and Salem al-Hazmi [two hijackers], did not have knowledge or fiqh ["jurisprudence"] in the popular terms. But they have fiqh in the tradition of Mohammed, peace be upon him, ... which is self-sacrifice for "No God but God" [Islam]. The sermons they gave in New York and Washington, made the whole world hear—the Arabs, the non-Arabs, the Indians, the Chinese—and are worth much more than millions of books and cassettes and pamphlets [promoting Islam]. Maybe you have heard, but I heard it myself on the radio, and the brothers in Europe have told me, that at one of the Islamic centers in Holland, the number of those who have converted to Islam after the strikes, in the first few days after [the attacks], is greater than all those who converted in the last eleven years.*

> **All those present:** *Glory be to God.*

> **bin Laden:** *I heard one person on the radio, the Voice of America, [unclear], they have an Islamic school there. He said, "We don't have enough time nor can we provide*

*enough books for all the people who are asking for books
and education about Islam." They [Westerners] are saying that these people must believe in something truly
great to push them to do such a thing. This will...And
we ask God to accept our deeds.*

al-Harbi: *I mean, how long people have been urging
people, "mobilize, make jihad. The infidels are your
enemy. The infidels, the hypocrites, they will rule you."
Sure, some did mobilize for jihad, but God, now that this
event has taken place; they [are] mobilizing by the hundreds. By the hundreds, Glory be to God...Like you said,
if people kept on inciting people to mobilize for decades, it
would not be the same push as this event. This has clearly
defined the [fault] lines, thank God.*

bin Laden: *We planned and made calculations. We sat
and estimated the casualties of the enemy. We figured that
[the casualties would be] the number of passengers in the
four planes; those will die. In regard to the towers, we assumed the [casualties] would be the number of people in
the three or four floors that the planes crash into. That
was all we estimated. I was the most optimistic. Due to
the nature of my profession and work [construction], I figured the fuel in the plane would raise the temperature of
the steel to the point that becomes red and almost loses its
properties. So if the plane hits the building here [gestures
with hands], the portion of the building above will collapse. That was the most we expected; that the floors above
the point of entry would fall...*

Al-Harbi told his hosts how elated Saudi Arabians were by the "hit."

> I was saying to myself, "the guys over there [in Afghan-istan] must be indescribably happy." Just see our hap-piness here [in Saudi Arabia]. The congratulations I received by phone that day, oh my God. My mother was busy all day just taking congratulatory calls for me all day, and relaying them to me. All day. So I figured that they must be even happier [in Afghanistan]... Undoubt-edly, Sheikh, this is a great victory. We ask God to bless this [Islamic] nation and give it honor and make it a na-tion of pioneers, because it deserves that... Thank God, it [America] is out of its hole. This is the second Ad [an Ara-bian tribe destroyed by God, which the Qur'anic equiva-lent of Sodom and Gomorrah]. God destroyed Ad, and He will destroy the second Ad [America], God willing. But he will destroy it at the hands of the believers. God willing, the true believers that have been waging jihad for a quar-ter of a century—the steadfast. I swear by the one and only God that I am living in a pleasure and happiness that I have not felt in a long time. Everyone is. I remember the words I am not qualified to speak.

The greatest act of jihad would be to "make jihad against the criminal nation," al-Harbi concluded, quoting the "Sheikh of Islam" Ibn Taimiyyah, the medieval scholar (1263–1328), and referring to the United States.

Al-Harbi was in the minority, however. Most jihadists

were opposed to what bin Laden had done, some even within his own wing of the movement. Abu al-Walid al-Masri, a senior member of the Al Qaeda shura, had been a leading theoretician of the network and participated in its most significant decisions. He was among the most senior of the Arab Afghans to break with bin Laden over September 11, and take his grievances public. He publicly lambasted bin Laden's "catastrophic leadership" and his underestimation of American willpower. Based in Qandahar, he supervised *The Islamic Principality*, a newsletter regarded as the mouthpiece of Mullah Omar. He also published a series of articles entitled *The Story of the Arab Afghans: From the Entry to Afghanistan to the Final Exodus with Taliban* in *Asharq al-Awsat*.

Abu al-Walid had worked closely with both Mullah Omar and bin Laden. He paints a dark portrait of bin Laden as an autocrat, running Al Qaeda as he might a tribal fiefdom. Bin Laden had ignored the advice of many of the hawks and doves around him, wrongly assuming the United States to be much weaker than it proved to be. Bin Laden had thought that the United States would retreat after two or three engagements, having based his assessment on the U.S. Marines "fleeing" Lebanon in 1983 and on what happened in Somalia in the 1990s, when U.S. forces left in a "shameful disarray and indecorous haste." But as Abu al-Walid notes, after September 11 matters "took an opposite turn compared to what bin Laden had imagined. Instead of buckling under his three painful blows, America retaliated and destroyed both the Taliban and Al Qaeda."

Al Qaeda members knew better than to challenge bin Laden, Abu al-Walid revealed. "You are the emir, do as you please!" he reported them as telling their leader. That attitude, the bin Laden aide wrote, turned out not only to be wrong but dangerous. "It encourages recklessness and causes disorganization, charateristics that are unsuitable for this existential battle in which we confront the greatest force in the world, U.S.A. It is therefore necessary to consider the real nature and the size of this battle as well as to prepare for it in a way that takes into account its danger and, consequently, to mobilize the mujahedeen and the Muslim masses for an extended, long-term battle that requires great sacrifices. It was necessary to prepare for the worst scenario that could come of this battle rather than dreaming of an easy victory."

By stifling internal debate and underestimating the enemy, bin Laden was personally responsible for the defeat, rendering Al Qaeda's final years in Afghanistan "a tragic example of an Islamic movement managed by a catastrophic leadership. Everyone knew that [bin Laden] was leading them to the abyss and even leading the entire country to utter destruction, but they continued to bend to his will and take his orders with suicidal submission." At certain points Abu al-Walid takes ad hominem shots at bin Laden, pointing to his "extreme infatuation" or "crazy attraction" with the international media. Bin Laden basked in the limelight and exaggerated his strength and capabilities. It is no wonder, Abu al-Walid tells us, that bin Laden entangled the Taliban in regional and international conflicts against its will and brought about the destruction of the Islamic emirate; Afghanistan was lost because of bin Laden's reckless conduct culminating in the attacks on the United States.

What seemed to fuel Abu al-Walid's anger was that bin Laden "was not even aware of the scope of the battle in which he opted to fight, or was forced into fighting. Therefore," he concluded, bin Laden "lacked the correct perception and was not qualified to lead." He cited an old Arab proverb to explain the catastrophe: "Those who work without knowledge will damage more than they can fix and those who walk quickly on the wrong path will only distance themselves from their goal." Abu al-Walid could not forgive bin Laden for abusing the hospitality of his hosts, the Taliban, by bringing destruction down upon them.

The Afghanistan disaster demonstrated one essential principle, according to Abu al-Walid: "The fundamentalists finally discovered from their experience in Afghanistan something of which they remained oblivious for several centuries: that absolute individual authority is a hopelessly defective form of leadership, an obsolete way of organization that will end in nothing but defeat." His verdict is damning: bin Laden's authoritarian style of leadership was responsible for pitting jihadists against America, which, in his opinion, is "beyond present capabilities of the whole [Islamist] movement."

Abu al-Walid argued that things went deeper than that. What happened in Afghanistan demonstrated the very intellectual bankruptcy of the jihadist project: "It may be that the Islamic movement had already suffered from an intellectual as well as an organizational defeat before it even had started its battle against America (otherwise known as the Great Satan). Jihad is a bigger and a more serious issue that should not be left to the jihadist groups alone. Jihad is more than just an armed battle."

While Abu al-Walid saw September 11 as a calamity, Abu-Jandal saw it as a golden opportunity to tilt the global balance of power in Muslims' favor. "Yes, America has primarily served Sheikh Osama bin Laden and Al Qaeda organization," he said in his interview with *Al-Quds al-Arabi*. "When Sheikh Osama attacked America, he wanted to expose it to the Islamic world. He also sought to expose its evil. This is what really happened. After the September 11 events, America stopped trusting any of its cadres [the CIA and the State Department]. It stopped trusting even its officers and officials. It is incorrect to say that Washington used Al Qaeda to swoop down on the Islamic world. America entered Iraq and before it Afghanistan as part of a long-term plan aimed at occupying the whole region. It wants to revive the history of the 1930s, 1940s, and 1970s. It wants to restore the French, British, and Italian colonialism to the region. America wants to occupy the whole region. Sheikh Osama and Al Qaeda caused America indigestion in swallowing up our Islamic world and prevented it from implementing its terrible scheme."

In the Muslim world today, you have jihadists who side with Abu-Jandal, but Abu al-Walid's withering criticism of bin Laden has been echoed by others, including the Egyptian al-Jama'a al-Islamiya. Whereas at the height of its strength in

2001, Al Qaeda membership never exceeded 10,000 people, al-Jama'a fielded more than one hundred thousand fighters in the 1990s. Of all the Islamists, al-Jama'a senior leaders, most of whom have been in prison in Egypt since the 1980s and 1990s, presented the most comprehensive critique of bin Laden's global jihad. Since early 2002 they have released eight manuscripts in Arabic, two of which deal specifically with the September 11 attacks. These are vital historical documents, shedding light on the thinking of the biggest and most influential jihadist organization in the region. Neither of the manuscripts has been translated into English, and thus not received the attention they deserve. The first, authored by Mohammed Essam Derbala and reviewed and approved by the entire leadership, is titled *Al Qaeda Strategy: Mistakes and Dangers,* and the other, authored by Nageh Abdullah Ibrahim, is titled *Islam and the Challenges of the Twenty-First Century.* Both were serialized in *Asharq Al-Awsat.*

Derbala, one of the leaders of al-Jama'a, is currently serving a life sentence in prison for his role in the 1981 Sadat assassination. He drew on religious texts to show that Al Qaeda's attacks violated Islamic law, which "bans killing civilians" of any religion or nationality. Derbala and his associates denounced Al Qaeda for preaching that American and Muslim interests would never meet and that "the enmity is deeply embedded and the clash is inevitable." They cited several cases in the 1990s when the United States helped to resolve international conflicts, with results that benefited Muslims: American military and financial assistance in the Afghan war tipped the balance in favor of the mujahedeen against the Russian occupiers; from 1990 to 1991 the United States helped Kuwait and Saudi Arabia expel Iraqi forces from Kuwait; in 1995 American military intervention put a stop to the persecution and massacre of Bosnian Muslims by Serbs; and in 1999 the United States led a NATO military campaign to force Serbia to end ethnic cleansing in Kosovo.

All these examples show clearly, al-Jama's senior leaders assert, that American and Muslim interests can and do meet; history has proven that there is nothing inevitable about a clash of cultures between Islam and the West because Islam is a universal religion, fully integrated with other civilizations. They reprimand bin Laden for advocating war between *dar al-iman*, "house of belief," and *dar al-kufr*, "house of unbelief." This is misguided and based upon a misreading of the ummah's capabilities. "The question is, where are the priorities? Where are the capabilities that allow for all of that?" they ask. Instead of embarking on what they regard as a blind suicidal approach, al-Jama'a leaders call for engagement with the West based on mutual respect and peaceful coexistence.

Derbala views the last sixty years of American policies toward Arabs and Muslims as, on the whole, negative and oppressive. Nonetheless, he rejects armed confrontation as a useful solution. Instead of deterring the United States, he maintains, "Al Qaeda boosted the anti-Islamic wave in America and the West" and widened the cultural gap between Muslims and Westerners. Derbala rejects bin Laden's and Zawahiri's assertion that the West is waging a crusade against Islam and Muslims: "Some claim that there is a crusader war led by America against Islam. However, the majority of Muslims reject the existence of crusader wars." Religious motives may influence American policy toward Muslim nations, he adds, "but these are not crusader wars." Rather, "interests remain the official religion of America, and those interests determine its international relations."

Thus, "Al Qaeda's policy helped crusading and anti-Muslim forces in America and the West to advocate a total war against Islam." If Al Qaeda proved capable of mastering anything, it was the art of making enemies rather than following Prophet Mohammed's example of neutralizing enemies. Al Qaeda was trying to ignite a clash of civilizations without possessing the means to wage—let alone prevail in—a global struggle. Echoing Abu al-Walid, Derbala insists that jihad must not be waged without an honest assessment of costs, benefits, and capabilities. "Al Qaeda has to understand that jihad is only one of the Muslims' duties. Jihad is a means, not an end." Making jihad for the sake of jihad, as Al Qaeda has done, is counterproductive because it produces the opposite of the desired results—the downfall of the Taliban regime and the slaughter of thousands of young Muslims. Surely, the ummah is much worse off now, Derbala points out, because of Al Qaeda's foolish and reckless conduct.

What I find most fascinating about the document is the way in which Derbala uses the very terms bin Laden and Zawahiri adopted to justify their jihad to point to its illegitimacy. He accuses them of violating the shariah itself, waging illegitimate jihad by superimposing their own views over those of the Prophet. He comes close to calling the Al Qaeda chiefs apostates. Still, bin Laden and Zawahiri could cut their losses, Derbala concludes, if they halt their jihad and concede their errors; otherwise, they will meet a fate similar to that of the Algerian Armed Islamic Group (GIA), a criminal gang that forsook Islam and met defeat at the end of the 1990s.

The main author of al-Jama'a's second manifesto, Nageh Abdullah Ibrahim, also sentenced for life for Sadat's assassination, writes that Muslims must relinquish myths maintained by extremists like himself for decades. According to Ibrahim, September 11 and its reverberations exposed the need for Muslims to face reality head-on and make difficult decisions if they want to catch up with the rest of humanity: Muslims can no longer afford to postpone reforms in a world whose social, political, and economic interactions are evolving quickly, leaving them further and further behind: "Standing still would mean suicide," he acknowledges.

A real renewal of Islamic thought, Ibrahim posits, would enrich the education of young Muslims and make them less vulnerable to easy conspiracy theories, such as those that got spun around Septermber 11. "Conspiracy theory retards the Arab and Muslim mind, restricting its ability to rationally resolve problems." Ibrahim laments the fact that instead of viewing foreign affairs as based on state interests and power relations, Arabs have acquired a conspiratorial lens that scapegoats the West for "all of our tragedies and neglects our own strategic errors." Those strategic errors—not the

West—he writes, are the real villains behind the decline of the ummah, and he concludes by saying that Islamists and nationalists are equally responsible for conspiracy mongering and leading young Muslims astray.

Ibrahim, Derbala, and the others cited their own experience fighting the Egyptian government to show the pitfalls of engaging in jihad without considering conditions at home and abroad; jihad not only failed to achieve their goals, but, more important, it lost them public support. Their error resulted, Ibrahim stressed, from forgetting "that armed struggle or jihad was never an end in itself, and Islam did not legislate fighting for the sake of fighting or jihad for the sake of jihad." Jihad is only one of Islam's duties; Muslims must not overlook other choices such as *al-solh*, "peacemaking," practiced by the Prophet Mohammed throughout his life. By neglecting al-solh as a strategic choice, all jihadists—and they include themselves in this estimation—made grave mistakes that endangered their movement's very survival.

Bin Laden, Ibrahim argues, violated the fundamental precepts of Islamic wisdom, which requires that faith be yoked to strength, justice, and tolerance. Even as he preached the value of piety and faith, bin Laden should have listened to his own internal counsel and tried to understand his adversaries. In the end he fell victim to hubris. Ibrahim points out that bin Laden "wants to fight America on September 11, the Russians in Chechnya, India in Kashmir, as well as carry out military operations in Muslim lands in Saudi Arabia, Yemen, Morocco, Indonesia, and elsewhere." Because it lost touch with reality, rationality, and the essence of what Islam is all about, Al Qaeda caused the downfall of two Muslim regimes—in Kabul and in Baghdad. Ibrahim sees little difference between bin Laden's Al Qaeda and Saddam Hussein's dictatorship: one leader destroyed his own network, the other destroyed the Iraqi state.

Derbala, Ibrahim, and their imprisoned colleagues condemn bin Laden and Zawahiri's religious justification for attacking the Americans, reminding them that Islam has always practiced—not just taught—peaceful coexistence as a permanent way of life. Religious coexistence is a strategic, not a tactical, goal in Islam, particularly when Muslims migrate to foreign lands and are welcomed by native inhabitants. What makes the crime of the September 11's suicide bombers uniquely un-Islamic, Ibrahim writes, is that the U.S. government had admitted them as guests. This was a betrayal of the most fundamental spiritual obligation, the one practiced in shops, cafés, and homes throughout the Arab world. Had the bombers read the Sunnah (containing the deeds of the Prophet; the second source of Islam after the Qur'an), they would have respected peaceful coexistence.

Speaking from his prison cell, Karam Zuhdi, the emir of al-Jama'a, gave a series of interviews to the Egyptian weekly magazine *Al-Mussawar* and to *Asharq Al-Awsat* in which he offered an even more pointed critique of September 11 and Al Qaeda. According to Zuhdi, bin Laden and Zawahiri did not understand that the bipolar American-Russian rivalry had been replaced by a unipolar U.S.-dominated system. Failing to recognize America's global supremacy, Al Qaeda dragged the ummah into a confrontation it neither desired nor had the capability to pursue. Refusing to accept this failure, bin Laden became "obsessed with killing Americans, Christians, and crusaders without distinctions."

What is the alternative to all this mayhem?" the imprisoned leaders asks. The United States should of course pursue a more just foreign policy, and Muslim states should empower their citizens by extending freedom and democracy to everyone. Jihad should only be activated against foreign aggressors and occupiers. One cannot blame the decline of the ummah on the enemies of Islam, as bin Laden, Zawahiri, and their cohorts do.

Al-Jama'a's critique of September 11 is all the more powerful because the credibility and legitimacy of its leaders cannot be questioned, even by Al Qaeda. Zuhdi, Derbala, Ibrahim, Osama Hafez, Assem Abdel-Maged, and the rest who signed and blessed the two documents from which I have quoted extensively, were the founding fathers of a major wing of the jihadist movement. While students at Asyut University in the late 1970s, they published one of the first manifestos of violent jihad, entitled *Chapters from the Charter of Islamic Political Action*. They paid their dues in blood and sweat and have languished in prison for decades. Like Kamal, they were the pioneers of the jihadist journey.

Nor are they alone. Other Islamist leaders have condemned Al Qaeda's internationalization of jihad, notably Montasser Zayat, who in the early 1980s served time in prison with Zawahiri and Kamal for involvement in the Sadat assassination. An attorney who defends Egyptian Islamists, Zayat has been privy to the inner circle of jihadists in Egypt and other Middle Eastern countries. He published two personal memoirs, which as I indicated earlier, reveal in harrowing and intimate detail the strength of the bond between imprisoned jihadists. Yet both books are also highly critical of Al Qaeda's attack on the United States. After Zawahiri's post–September 11 memoir, *Knights Under the Prophet's Banner,* raised doubts about Zayat's loyalty to the cause, Zayat also published his diaries to clear his name and prove his jihadist credentials. Zawahiri had accused his old comrade of having suspicious connections with Egyptian security officials.

The split between Zawahiri and Zayat had its roots in Zayat's having advocated for al-Jama'a's 1997 ceasefire among jihadists, including Zawahiri's Tanzim al-Jihad. Zayat became the messenger of the peace initiative, which had first been proposed by al-Jama'a's imprisoned leaders, as they struggled to end the state of war between Islamists and the Egyptian government.

I interviewed Zayat in his Cairo law office in 1999 and 2000. Though careful and measured in his speech, he was not an especially modest man; he seemed to enjoy the limelight. Most of our conversations revolved around his clients' peace proposal and the difficulties he was facing from the rejectionists in the government and the Zawahiri camp. Both sides were bent on undermining the ceasefire he had helped broker. He told me that he was prepared to show me new al-Jama'a manifestos, calling on their foot soldiers to lay down their arms and end their insurgency, if I would publicize them. I politely declined. Zayat's objective was to tell the world—in a way that the Egyptian government could not ignore—that a profound shift had taken place within the jihadist movement(s). While Zayat acknowledged that Zawahiri and his militant allies opposed the ceasefire initiative, at that point he was still circumspect in his criticism of his former associates.

But after September 11 Zayat was less restrained about Zawahiri and Al Qaeda. His diaries portray Zawahiri as a reckless opportunist with no moral scruples. Drawing on conversations with hundreds of Islamists and jihadists over the previous twenty years, he reproaches Zawahiri for opening a second front against a far more superior enemy. How could Zawahiri commit such a fatal strategic error and disregard the primacy of establishing an Islamic state in Egypt? The answer, in his view, is simple: egoism. Zayat also makes no effort to mask his contempt of bin Laden, though he does acknowledge that bin Laden was more consistent than Zawahiri, for all along he had been struggling to expel the Americans from the Persian Gulf, particularly Saudi Arabia.

Zawahiri, on the other, was nothing but an overambitious, vain, irresponsible tactician who cared less about the future of Tanzim al-Jihad than about magnifying his own image and status through an unholy alliance with bin Laden. And for what? Zawahiri turned the Tanzim from an organization "aimed at building an Islamic state in Egypt into a branch within Al Qaeda, subordinating a well-established organization to a new, experimental one—Al Qaeda—which subsequently caused considerable harm to Islamist groups and activists throughout the world."

To reassure his Islamist friends, Zayat stresses his loathing of American foreign policy, which he sees as hostile to Arabs in particular and Islam in general. Resistance to American imperialism remains for Zayat a religious duty. Yet any effective strategy of resistance must be informed by costs, benefits, and the balance of power whereas September 11 was driven by a simplistic desire for revenge. The consequences have proven disastrous for the entire Islamist movement. All bin Laden and Zawahiri accomplished, Zayat and his allies believe, was to unify the international community against what had been a vigorous return to Islamic fundamentals. Who would have thought, Zayat asks, that European governments, which had historically granted political asylum to radical and militant Islamists, would not only no longer take them in, but would repatriate them to their home countries to face trial, torture, and persecution?

Were the American and European counterattack merely military, that would have been bad enough. But Zayat believes that in their rage Westerners have become bent on total elimination of the Islamist movement as a military and political force. The movement's ability to withstand the American storm, Zayat maintains, will depend on the willingness of its leaders to reflect critically on what went wrong; they must take stock and quickly repair the damage inflicted by Zawahiri and bin Laden, who forced the jihadist caravan off track. In short, if they are to survive, Islamists must construct a long-term strategy to resist the onslaught by the new imperial power.

Zayat does not lay out a blueprint for militant Islamism.

His conclusions are vague. Yet after the publication of his critical memoir, no knowledgeable observer could deny that he has revealed the fault lines among jihadists. One hard-line cleric, Omar Mahmoud Abu Omar, also known as Abu Qatada, attacked Zayat for being motivated by revenge against Zawahiri, dismissing his memoir as a "deviant case" and "evil analysis." Abu Qatada, a Palestinian preacher who has lived in Britain since 1993, is sometimes called bin Laden's spiritual ambassador in Europe. He is currently under house arrest under a British law introduced after September 11, permitting the detention without trial of foreigners deemed a danger to national security.

The majority of Islamists and jihadists, however, have echoed Zayat's view of September 11: that it was a catastrophic blunder. In his own diaries, serialized in *Al Hayat*, Hani al-Sibai, an alleged leader of the Jihad Group who resides in exile in Britain (the Egyptian government has sentenced him to death), is bluntly critical of Al Qaeda and September 11. Since its birth in 1998, he writes, the global jihad movement led by Al Qaeda has proved disastrous to the Islamist movement and the ummah alike. Like Zayat, Sibai calls the decision to shift operational priorities and attack the United States unwise, based neither on rational analysis nor on consultation with the rank and file.

Except for al-Jama'a's chiefs, who stress moral and ethical factors in their condemnation of September 11, most Islamists' criticisms are based on utilitarian and pragmatic considerations. In his rebuttal of Al Qaeda and September 11, Osama Rushdi, who was in charge of al-Jama'a's media committee and a senior member of its consultative council, comes close to coupling the moral with the political. In several interviews with the Arab media, Rushdi makes the point that although Al Qaeda justified its attacks on the United States in religious terms, those terms had nothing to do with Islam. Islam does not sanction killing civilians or violating legal and moral percepts, he stresses, because that would threaten international harmony and coexistence. Rushdi makes it clear that he opposes Al Qaeda's internationalization of jihad and its so-called blessed terrorism. If Al Qaeda members truly respect the rules established by the shariah for pursuing jihad, he insists, they should reflect on their errors and correct them before it is too late.

Criticizing American foreign policy is easy, Rushdi acknowledges. He directly addresses the Al Qaeda jihadists whom he holds accountable for the current crisis: "Does hostility to America justify utilizing all means to attack it and harm its citizens regardless of their legitimacy and the inherent benefits and costs? Do the ends justify the means in this struggle, or should the means be as justifiable as the end?" The greatest threat facing the jihadist movement, he warns his former associates, lies in self-inflicted wounds. For too long jihadists and Islamists have neglected building institutions, preferring to grant "blind obedience to the charismatic leader who surprises his companions with abrupt decisions to the extent that they find out about them in newspapers"— a direct reference to the fateful decisions taken by bin Laden and Zawahiri.

I called Kamal and asked him whether he felt as strongly

as al-Jama'a's leaders and others about September 11 and bin Laden's recklessness. He took a deep breath before launching into a discursive response on the historical context of the attack, and that it was a tit for tat against hostile American policies. I said I had expected him to be more vehemently critical of Al Qaeda's September 11 than his counterparts; I reminded him that when we first met he had hardly anything positive to say about the Afghan generation, one that substituted military muscle for vision.

"Listen," Kamal retorted over the phone, "I feel pity and empathy for al-shabab Al Qaeda. They were used as ammunition and fuel by America and its ruling Muslim allies to defeat the Soviet communists. But once victory was had, al-shabab were treated like criminals by their own governments and America; they felt that they were used and abused by America. Al-shabab had hoped to enjoy the fruits of their victory over communist Russia and be rewarded by playing leadership roles in their countries. Instead, what occurred was the exact opposite; many could not return home because they were blacklisted by local authorities as security risks. Al-shabab held America responsible for their predicament because it controlled the political order in the region; in their eyes, America replaced communist Russia as the incarnate devil." In his justification of September 11, Osama bin Laden cited America's hypocrisy regarding the Afghan Arabs. "History reveals that America supported everyone who waged jihad and fought against Russia, but when God sanctioned these Arab mujahedeen to help those poor innocent women and children in Palestine, America became angry and turned its back, betraying all those who had fought in Afghanistan."

"What does the killing of thousands of American civilians have to do with how al-shabab felt about U.S. policy?" I asked "Do not misunderstand me," Kamal quickly added. "I did not support September 11 and the massacre of civilians. But there is a bigger point to be made. The September explosions could not be understood without discussing American policies and practices that enraged al-shabab and led them on their fateful journey," he said, choosing his words carefully. It was not just that American politicians violated an unwritten understanding with al-shabab but also turned against them with a vengeance. After the defeat of the Soviets in Afghanistan, he said, America viewed the mujahedeen as threats to its dominance and decided to clip their wings; the collapse of the Soviet empire also caused American strategists to look for new enemies and by the early 1990s they found Islam and the Islamists. Kamal advised me to reread the political literature of that period, especially the articles "The Clash of Civilizations" by Samuel Huntington, and "The End of History" by Francis Fukuyama (both essays were turned into books and triggered national and international debate).

America went further than delineating Islamists as the enemy, Kamal told me. It actively assisted local regimes in hunting down leaders of the movement: "In the 1990s America declared war on Islamists and enabled Muslim rulers to inflict blows on them." The 1991 stationing of U.S. troops in Saudi Arabia, the birthplace of the Prophet, was also a big emotional issue for Osama bin Laden and his comrades, he added; a provocative act that enraged al-shabab. Many people do not know this history, Kamal went on, and overlook the fact that America went to war against Islamists first.

"I still do not understand what you are trying to say," I retorted.

"The September bombings were retaliation against U.S. policies that did a lot of harm to al-shabab," Kamal shot back. "It was a desperate act to punish America and teach it a lesson. Of course, the attack on America was counterproductive to the Islamist movements and the ummah but you must understand the historical context that brought us to that point. Al Qaeda organization does not speak for Islamists; it is a marginal current within the movement. But al-shabab were fed up with American arrogance."

I asked Kamal why he viewed September 11 slightly differently than al-Jama'a's imprisoned chiefs and other radical Islamists like Rushdi, Sibai, and Zayat. "We all agree that the September bombings have had a negative return but we have different tones and different ways of expressing our opinion," he acknowledged. "While I fully agree that the attack on America did not serve the cause, I also view it in a broader historical sense—the culmination of the struggle between the United States and Al Qaeda organization. The war between them broke out in the early 1990s, not on September 11, 2001." "Did Islamists benefit in any way by September 11," I asked. "No," he replied. "The September bombings harmed the Islamist movement and dragged the ummah into an uneven fight."

The predominant public response to September 11 by Kamal's generation was to condemn Al Qaeda. Privately, they confided their fury with bin Laden and Zawahiri. The general realignment within the jihadist family has turned decidedly against the global jihad. If the movement's major figures—Abu al-Walid, Sibai, Zuhdi, Rushdi, Derbala, Ibrahim, Zayat, and many, many others who have avoided making public statements—do not follow Al Qaeda's lead, who would? Where will Al Qaeda find new recruits?

Instead of closing ranks against "the enemies of Islam," as bin Laden and Zawahiri had hoped, September 11 put to rest any possibility of bridging the gulf between local and global jihadists. Al Qaeda is unquestionnably the real loser, for it desperately needs loyal allies and revolutionary legitimacy; its supposed natural partners not only deny it that recognition but attack it relentlessly. Zawahiri's and bin Laden's most recent pronouncements confirm that their appeals to Muslims to rise up and join the fight have largely fallen on deaf ears. Neither the ummah nor the army of deactivated local jihadists has been willing to fight alongside Al Qaeda, however much they empathize with its grievances against the international order and American foreign policy in particular.

Until, that is, the second war in Iraq.

VI

JIHAD DIASPORA

AT THE PEAK of rush hour on the morning of July 7, 2005, four Muslim men in their late teens and early twenties carried out a series of coordinated bombings on three London underground trains. A fourth bomb exploded an hour later on a double-decker bus in central London. Fifty-six people died, including the four bombers, later identified as three British-born men of Pakistani descent and one of Jamaican ancestry. Seven hundred more were injured. The attackers, who all lived in the north of Britain, came from middle-class families. For weeks after, Britain was gripped with lingering questions: Why would second-generation British Muslims, whose immigrant parents had climbed the social ladder, reject the country of their birth? What had led them on their suicide journey? On most people's minds was the question whether the attacks truly carried "all the hallmarks" of Al Qaeda, as British officials said immediately afterward.

Eight weeks later, in a videotape broadcast on Al Jazeera, the Arabic satellite network, Mohammed Sidique Khan, a thirty-year-old resident of the English city of Leeds, who had by now been identified as the ringleader of the group, offered—in English—his reasons for the attack. "I am going to keep this short and to the point because it's all been said before by far more eloquent people than me," said Khan, who had a trimmed beard and was wearing a red-and-white checkered keffiyeh (headdress) and a dark jacket. He was leaning against a wall on which hung an ornate carpet. "But our words have no impact upon you. Therefore I am going to talk to you in a language that you understand. Our words are dead until we give them life with our blood.

"I'm sure by now the media has painted a suitable picture of me. This predictable propaganda machine will naturally try to put a spin on it to suit the government and to scare the masses into conforming to their power and wealth-obsessed agendas. I and thousands like me are forsaking everything for what we believe," Khan said, pointing his finger at the camera. "Our driving motivation doesn't come from tangible commodities that this world has to offer.

"This is how our ethical stances are dictated: You democratically elected governments continuously perpetuate atrocities against my people, and your support of them makes you directly responsible, just as I am directly responsible for protecting and avenging my Muslim brothers and sisters." He made a solemn pledge that "Until we feel security, you will be our target. Until you stop the bombing, gassing, imprisonment, and torture of my people, we will not stop this fight. We are at war and I am a soldier. Now you, too, will taste the reality of this situation," he concluded.

The video included an appearance by Zawahiri, in which he praised the July 7 bombings as "the blessed London battle, which came as a slap to the face of tyrannical, crusader British arrogance... Like its glorious predecessors in New York, Washington, and Madrid, this blessed battle has taken the war to the enemies' land." Zawahiri spoke in Arabic. He was dressed in a black turban and white robes. An automatic weapon was leaning against the wall beside him. While the two men did not appear together on the tape, Zawahiri was clearly trying to lay claim to the London bombing, which he said were a direct response to Britain's participation in the U.S.-led invasion and occupation of Iraq and its rejection of a truce offered by bin Laden. After the March 2004 train bombings that killed 191 people in Madrid, Spain, bin Laden had proposed to European countries a three-month ceasefire during which to consider his demands for a withdrawal of their troops from Iraq and Afghanistan. "Oh, nations of the Christian alliance, we have warned you before. So taste some of what you have made us taste," Zawahiri, now Al Qaeda's public face, said. "We will respond in kind to all those who took part in the aggression on Iraq, Afghanistan, and Palestine. Just as they made rivers of blood flow in our countries, we will make volcanoes of anger erupt in their countries." Addressing Britons, Zawahiri added: "Blair not only disregards the millions of people in Iraq and

Afghanistan, but he does not care about you, as he sends you to the inferno in Iraq and exposes you to death in your land because of his crusader war against Islam."

Both Zawahiri and the British government went to work to convince their audiences that Al Qaeda had carried out the July 7 bombings. Immediately after the attack, British officials not only swiftly blamed Al Qaeda but disputed reports that the bombings had been inspired by generalized Muslim anger at the decision by Prime Minister Tony Blair to commit troops to the American-led invasion and occupation of Iraq. For its part, Al Qaeda wanted to reassure its base and sympathizers that it was still capable of taking the war to Western capitals, that it was still the vanguard, leader, and trailblazer of the ummah—the only one to exact revenge on its enemies.

We know very little about Mohammed Sidique Khan and his fellow suicide bombers. Unlike earlier generations, these jihadists left behind no extensive testaments or diaries apart from Khan's videotape, which doesn't do much to explain why a small contingent of second- and third-generation Europeans, mostly middle class and educated, had responded to Al Qaeda's call. Speaking in a heavy Yorkshire accent, Khan praised "our beloved Osama bin Laden." Khan and his friends apparently saw themselves as foot soldiers, doing battle on behalf of an ummah under attack by Western nations.

They were not the first, and likely won't be the last, suicide bombers born and raised in the heart of Europe. In 2003, two natives of Derby (one born there, one a childhood immigrant), Asif Mohammed Hanif and Omar Khan Sharif, outraged by Israel's oppression of the Palestinians, traveled to a Tel Aviv discotheque and carried out a suicide mission, killing several people. Unlike their parents, whose practice of Islam was less politicized, Khan, Hanif, Sharif, and others felt that the struggle of their coreligionists in distant Muslim lands such as Kashmir, Chechnya, and Iraq was their own. However, the roots of their revolt are local, and can be found in their cultural and social segregation from mainstream European societies. Disenfranchised, unemployed, and alienated, living in ghettoized neighborhoods in London, Bradford, Leeds, Paris—as we saw in the November 2005 rioting in the suburbs—Lyons, Hamburg, Milan, Madrid, and other cities, they became easily drawn to self-righteously puritanical Sunni Salafism.

The migration of militancy into Muslim communities in Europe, though on a small scale, reveals how complex the journey of the jihadist is. On the surface, Kamal and Khan seem to be worlds apart, representing different generations and different socioeconomic and national backgrounds. While they each have traveled a separate road, they have arrived at a similar conclusion: the urge to sacrifice everything for a sacred view of the ummah—an imagined entity more appealing and enduring than their social and national milieu.

In November 2005 I called Kamal in Egypt to ask about the recent changes taking place within the jihadist family. It has been more than a decade since Kamal put aside armed struggle, advocating instead participation in the political process, yet he sounded gloomy. He told me that a global battle was raging between the worldwide jihad movement (still led by Al Qaeda) and other anti-U.S. forces, and a process of globalization spearheaded by America and its Western allies. "We are confronting new realities and new rules of the game, which are totally different from the previous rules." Whereas after September 11 and the debacle in Afghanistan there had been a stalemate between those focused on the far enemy and those on the near enemy, according to Kamal that balance was now shifting. More and more were turning away from jihad against local rulers and calling instead for an armed struggle against "America as an occupying power" of Muslim lands. "The fault lines have shifted from internal to external enemies, particularly against the new crusade," he told me.

The day we spoke was Eid al-Fitr, an Islamic holiday marking the end of Ramadan, when Muslims hold special congregational prayers and celebrate by visiting each other and feasting.

Kamal predicted that a new wave of jihadism would erupt whenever the ummah and its religious symbols and territories were perceived to be threatened by outside aggressors and when the shariah was not being implemented. Muslims feel they are facing an existential threat, a "new Christian crusade allied with Jewish fundamentalism" whose goal is to neutralize Islam by discrediting "positive Islamic values like jihad, fighting, martyrdom, and the idea that all Muslims belong to one ummah. The new crusade is targeting traditional Islam," he said.

"We all sense danger. I sense danger. The ordinary man on the street senses danger. America is trying to restructure our societies and ways of life, including the most sensitive aspect, our most important line of defense—the Muslim woman. Under the pretext of reforming and renewing religious texts, the new crusade wants to replace traditional Islam with Protestant Islam. People are scared and expect the worst," Kamal said with a sigh.

I wondered why the Muslim woman was now specifically under attack. What was happening to threaten this most sensitive aspect of Muslim culture? President Bush had recently sent Karen Hughes, his former communications director, on a goodwill mission to the Middle East, but I doubt that Hughes would have seemed threatening to someone like Kamal. Nonetheless Kamal seemed anxious about efforts by the United States to empower Muslim women and have them a play a bigger role in the public space. To Kamal and other radical Islamists, women's rights and empowerment are code words to weaken the ummah's immune system and make it vulnerable to decadent, Western influences; it is the most effective way to penetrate the Muslim social network and dismantle its values. The Muslim woman preserves the puritanical social order dear to Salafis like Kamal's heart. Once the West overruns this last line of defense, Kamal told me, the social network would fracture and collapse.

In 2004 Kamal commented to Al Jazeera that the next jihadist wave would be launched by resistance to the

"Christian-Jewish fundamentalism that is attempting to redraw the map, to make over the values, and religious, cultural, and historical identity of the Muslim world." The traditional polarity between the Qur'an championed by jihadists and the Muslim sultan (the local struggle) would be superseded by a confrontation against the American agenda. Kamal posited an alarming scenario in which the Islamic renewal, along with the Muslim populations and even Muslim governments, would repel the new crusade led by the Christian-Jewish religious right.

Kamal's outlook had darkened considerably since we first started speaking in 1999; his rhetoric had turned more militant. "Well, why are you shocked?" Kamal retorted when I pointed this out. "America and its allies are waging a total war against our most sacred values, our Arab-Islamic identity. Unlike other battles, the new fault lines go beyond the military into religion and culture."

The Prophet Mohammed caricatures published first in the Danish newspaper *Jyllands-Posten* and then in other European newspapers provide a case in point. After their publication a torrent of anger and rage swept Muslim countries from Indonesia to tiny Lebanon, at times turning violent. The reaction was an expression of an identity crisis—a hypersensitive state quick to perceive internal and external danger. According to most public-opinion surveys, ordinary Muslims as well as Islamists already fear that the "Christian West" is waging a war against them, one in which their spiritual values—their very Islamic identity—are being targeted. The Mohammed cartoons fed straight into that fear.

Kamal believes that Al Qaeda now acts as a transnational organization, an overarching structure for various groups opposing the "new American crusade." He cautioned me, however, against lumping the bombers in London and Madrid together with Al Qaeda. Khan and his teammates simply shared with Al Qaeda a "common ideological hostility against America" without maintaining organizational affiliation with the bin Laden network. "Al Qaeda may be considered as a paradigm and umbrella for all jihadist forces confronting America and its allies." Indeed, Kamal is of the opinion that Al Qaeda is a decentralized network of loosely linked factions and cells united in their struggle against American imperialism. His views are shared by Western scholars and intelligence services alike.

Kamal has noticed a new consensus among Islamists and jihadists of all colors that the United States had become the implacable enemy. "In our eyes, America tops the list of threats facing the Islamic nation." Like others, Kamal viewed the American war on terror as a war against Islam and Muslims. After the U.S. attacked the Taliban and Al Qaeda he wrote an article entitled "The Clash of Cultures Turns from a Slogan into a Strategy: America Declares World War Three against the Muslim World." "America's declaration of war on Afghanistan on October 6, 2001, was a declaration of war on Islam," he wrote. The war is not about September 11. "The truth is that World War Three by the Christian West on the Islamic East is a religious, cultural, and civilizational war designed to silence all those Muslims who still talk about holding fast to Islam in order to resist the dirty tide of the tyrannical Western civilization." Despite what Arab and Muslim governments want us to believe, Kamal tells his readers, "Taliban and Afghanistan represent in our estimation the first line of defense for the Muslim world in this new world war because our national security is one and the same."

After reading his recent commentaries, I asked Kamal about their tone. He acknowledged that they were angry, vocal, and confrontationalist. "I refuse to surrender to the arrogance of American power. I refuse to surrender as others have—like al-Jama'a al-Islamiya." He was shouting to make himself heard on the phone.

As we've seen, most jihadists and Islamists did not see Afghanistan as a "battle between right and wrong," as Kamal had put it. No religious figure lent his authority to repelling the U.S. invaders from Afghanistan, and a fatwa was issued declaring that American Muslims were obliged to serve in the armed forces of their country. Though Sheikh Yusuf Qardawi, who signed on to this fatwa, was and is highly critical of American policy toward Muslims, he was the first cleric to issue a fatwa that condemned Al Qaeda's "illegal jihad." Two days after September 11, he wrote that the attacks could not be justified on any grounds, not even "the American biased policy toward Israel."

Similarly, Sheikh Mohammed Sayyed Tantawi, the Grand Imam of Al-Azhar in Cairo, the oldest religious institution in the Islamic world, swiftly dismissed bin Laden's jihad credentials as "fraudulent" and warned young Muslims against heading Al Qaeda's call to fight in Afghanistan.

Militant Salafi clerics in Saudi Arabia and Yemen, who had previously been supportive of bin Laden, found themselves on the defensive and distanced themselves from Al Qaeda; they did not issue jihad fatwas.

In contrast to the war in Afghanistan, the 2003 U.S.-led invasion and occupation of Iraq triggered a torrent of angry responses by Islamists, ulemas, secular Muslims, and religious Muslims alike. Liberal Arab writers and artists, often maligned for their pro-Western stance, denounced America's "imperial hegemony." Adonis, a leading poet, expressed the sentiments of his generation:

> What is the goal of American policy toward the Arabs? It aims at keeping the Arabs behind history and without future. For every free and liberal thinker in the world, America's imperial policies engender anguish which transcends his private passion and pain. It creates civilizational agony for man and humanity.

Adonis's words express the anguish felt by those Arabs who while disposed, even drawn, to the West, were enraged by its militarism and aggression. Institutions and clerics urged Muslims to join in jihad with their Iraqi brethren and repel the American invaders. Al-Azhar issued a fatwa advising "all Muslims in the world to make jihad against invading American forces," and Tantawi, often accused by Salafi clerics and militant Islamists as being a pro-Western reformer, ruled that efforts to stop the American invasion were a "binding Islamic duty."

The fight against the Taliban in Afghanistan and the invasion of Iraq were being viewed differently. After coalition forces took Baghdad, the same Qardawi who had forcefully denounced bin Laden and his cohorts now accused the Bush administration of declaring war on Islam and behaving like "a god." There was no ambiguity in his stance. Fighting American troops was "legal jihad" and "death while defending Iraq a kind of martyrdom." In an interview on Al Jazeera, which has an audience of more than thirty million Arab viewers, Qardawi sanctioned attacks on Iraqi civilians who had commit the "crime" of assisting "the enemy." Earlier he was quoted as blessing the murder of American civilians in Iraq. Qardawi later denied that he had made the remarks and said that in any case his words were taken out of context, but the damage had been done. Qardawi and Tantawi's hardened stance on Iraq reflected a deepening cultural and religious divide between Muslim societies and the West.

In the eyes of Arabs today, Iraq competes with Palestine as the Muslim world's open wound. Both have become rallying cries for those who rant against the brutality of Americans and Israelis and the suffering of Iraqis and Palestinians. Kamal launched a frontal assault against what he labeled "the fascist clique that rules America," which saw the invasion as a divine mission. "What a strange travesty—'a criminal clique' intends to liberate [the Iraqi] people who had been oppressed by another criminal clique and usurp Iraq's resources...This is what happened in Iraq. The nationalist Baathist clique was replaced by a crusader-Zionist clique led by the conservative religious right that rules America." In a commentary appearing in the Islamist newspaper *Al-Shaab* after the fall of Baghdad in April 2003, Kamal lashed out angrily against "the liberation lie" being sold by the ruling cabal in Washington. America in Iraq is "an occupation power that aims at control, rape, and hegemony."

For Kamal and so many others, the American occupation in Iraq represents a carbon copy of the Jewish occupation in Palestine. "Both belong to one school of thought—domination—and the Iraqi and Palestinian people belong to another school of thought—resistance and rejection of occupation." Long before the escalation and intensification of the insurgency, Kamal predicted that the resistance to the American occupiers would be spearheaded by Iraqi Islamists—jihadists from within the Sunni Arab community in central Iraq and around Baghdad. Sunni Arabs, who represent about 20 percent of Iraq's population of twenty-three million (Kamal said they represent 35 percent, or a little over nine million people), feel targeted by what they see as plans to encourage sectarianism and destroy the Sunni Arab character of modern Iraq. He was not surprised that Arab jihadists were fighting alongside their Iraqi counterparts. "It is well known that America has many enemies in the world...and Al Qaeda organization might transfer its armed activities into the heart of Iraq...because it is a much more important strategic area than Afghanistan, and it is easier to bleed the Americans in Iraq than in Afghanistan."

I was stunned by how dramatically Kamal's views had evolved in the years I had known him. He now seemed in closer alignment with Zawahiri and other militants. "Yes,

my discourse on America's new crusade is closer to Al
Qaeda than to al-Jama'a al-Islamiya," he conceded. "We are
at war. America occupies a Muslim country in the heart of
the Arab world. I do not need to remind you of the historical and religious importance of Baghdad [the former seat
of the Islamic caliphate, the Muslim state that dissolved
with the fall of the Ottoman Empire after World War I] to the
ummah. America brought its armies, missionaries, and plans
to subjugate Arabs and wipe out their Islamic identity. How
do you expect me to respond, when my people are assaulted
by the greatest military power in the universe? In the 1990s
I called for dialogue with the West and the United States. But
not now! I believe that America and its fundamentalist ruling class are intrinsically hostile to Islam and Muslims. They
see us as evil and have plans to indoctrinate and brainwash
us and impose secularism on us. America's occupation of
Iraq left us with no choice but to defend ourselves and our
faith."

Kamal emphasized that he was referring to the U.S. political establishment and not to the American people. In fact,
he has told his Arab readers that Iraqi resistance exposed the
lies that Western politicians fed to their citizens. The American public were finally discovering that their leaders—"the
war merchants"—had misled them and were demanding
that they bring their sons back home. "I suspect that the
clock of history cannot be turned back." I asked Kamal why
he did not advocate appealing to the American people and
informing them about the injustices committed by their
leaders against Arabs and Muslims. Kamal replied that there

are many Americans who oppose their country's imperial policies with whom Islamists would like to cooperate. "We have seen hundreds of thousands of Americans protest their government's plans to invade Iraq; we have seen Americans oppose the abuse of Muslims' rights and liberties by their authorities. We can join hands together with these progressive currents and voices in America; those are our natural allies," Kamal reassured me.

But Kamal hurriedly added yet again that American values, such as human rights and the rule of law, had been hijacked by a fundamentalist clique. "Now religion plays an important factor in the making and execution of U.S. policy toward the Muslim world. The idea of manifest destiny and cultural superiority permeates U.S. official thinking and conduct. What happened to American laws?" Kamal inquired rhetorically. "They were slain," he replied, answering his own question. "Yes, we would like American citizens to reclaim their old liberal traditions and values. But we do not have the luxury to rely only on the goodwill of the American people; this would be futile. A cabal in foreign policy had convinced the American people that Muslims needed to be confronted and defeated."

His tone hardened when he mentioned President Bush's use of the term crusade to describe the American war on terror. "We were shocked to hear Bush utter the word *crusade* because he reminded Muslims of the Crusades against our ummah." Although Bush later apologized for using the term, the damage was done. "For us, *crusade* possesses a religious connotation—it is a religious war," Kamal told me.

"To be honest with you," Kamal confided, "after September 11 I discovered another America out there—imperial, racist with a superiority complex. It was a frightening discovery because I had thought that the liberal camp was dominant. I had not known that fundamentalists who believe in the clash of civilizations are deeply rooted in American society. This discovery came as a great shock to many of us who were calling for dialogue and engagement. The danger lies in that this fundamentalist camp is in charge of the most powerful nation in the world. It has a greater moral and legal responsibility than Al Qaeda organization, which is homeless and lawless. Since 2001 America has acted more like Al Qaeda than the nation of laws which it used to be. Look at the abuses of Muslims in American prisons all over the world. Look at the illegal arrests of many Muslims in America."

Islamic resistance, not military occupation or charity, would liberate Iraq and shape its future, Kamal stressed. It would frustrate and defeat the conservatives' efforts to impose their imperial designs on the Arab world. "Iraq has turned into America's new Vietnam...and it has proved that nation building by force is a big myth. The Iraqi resistance will not only determine the future of the country but also the future of the whole Arab region, its regimes, and institutions."

In my travels in the Middle East since the fall of Baghdad, I have heard similar arguments advanced by secular Arab nationalists and Islamists alike. *Al-Muqawama al-Iraqiya*, "Iraqi resistance," is being viewed as the only source of light in an otherwise dark tunnel. In March 2004, a pan-Arab nationalist leader, an Iraqi in his seventies, told me that al-Muqawama al-Iraqiya would tip the balance of power in the Arabs' favor, including in Palestine. Along with others, he had for decades believed in the centrality of the Palestinian cause to Arab political and cultural empowerment. Now he believed that the knights of Baghdad, not Jerusalem or Damascus, would restore Arab dignity.

I heard similar sentiments expressed during a conference called The Occupation of Iraq and its Repercussions Regionally and Internationally, which was organized by the Centre for Arab Unity Studies in Beirut in March 2004. Over one hundred and fifty opinion makers, activists, and politicians of all persuasions—nationalist, Islamist, and leftist—attended the four-day forum at the Crowne Plaza Hotel. I could find no one who supported the invasion and occupation of Iraq. Both Iraqi and Arab participants with whom I privately met were pinning their hopes on al-Muqawama al-Iraqiya. Their fervent belief was that the resistance should be supported with blood and treasure. "Al-Muqawama is our only hope to liberate Iraq and expel the American occupiers," declared an Iraqi Islamist from Falluja as we sat together at a café. A city of 300,000 that has been repeatedly assaulted and devastated by American forces, Falluja remains the heart of the Iraqi resistance. "Al-Muqawama is setting an example to be imitated throughout the Muslim world," he told me. An articulate and highly educated Iraqi from one

of the important Sunni Arab tribes, he asked me not to identify him by name in print. He feared for his life and safety,
he said. He might easily be arrested and tortured by U.S.
and Iraqi authorities. "We are facing the most powerful occupying empire in history and we will defeat it like we defeated the British empire before it," he told me with quiet
conviction.

I was puzzled by the vehemence of his views. Did not the
United States rid Iraq of the Saddam Hussein dictatorship? I
asked. His reply came out in a torrent. "Come with me to
Falluja and Baghdad and see for yourself how American soldiers treat ordinary Iraqis and tribal leaders. Come and see
how American soldiers and their trained Iraqi security men
violate the sanctity of our homes and bedrooms. Who gave
America the right to occupy a Muslim country? Was not
America Saddam's ally for many years in the 1980s? Bush
and his hard-line aides used Saddam as an excuse to occupy
Iraq and control our oil resources and divide us along sectarian lines. America will regret the day it set foot on our
Muslim lands. Al-Muqawama is bleeding the occupiers and
driving them to despair." He sounded both proud and
hopeful.

Many critics have pointed out that al-Muqawama has divided Iraqi communities further and killed thousands of
innocent civilians. I asked my Iraqi acquaintance if the insurgency had been infiltrated by terrorists. What role was
Al Qaeda playing, in particular bin Laden's emissary in Iraq,
the militant Jordanian Abu Musab al-Zarqawi? Notorious
for his ruthlessness, Zarqawi is thought to be behind the

beheadings of foreigners as well as a string of the most deadly suicide bombings. Recently, the U.S. put a $25 million bounty on his head—the same as on bin Laden himself. Has Al Qaeda turned Iraq into a major base of operations after being expelled from Afghanistan? Was al-Muqawama not playing into the hands of those in the Bush administration who were asserting that Iraq had become the central front in their war on terror?

"It is not true that Al Qaeda has a major presence within al-Muqawama," he retorted. "This is all American propaganda to justify the war. For your information, Arab [non-Iraqi] fighters do not exceed 5 percent of the mujahedeen. I know this fact from firsthand observation and from talking to people close to al-Muqawama. The Bush administration lumps all mujahedeen together with Al Qaeda fighters to discredit al-Muqawama and as a scare tactic at home. The big noise about Al Qaeda in Iraq is designed for American public consumption and to divide and conquer Iraqis. It is an old colonial trick." (According to American military commanders in Iraq, native Iraqis represent about 90 percent of all fighters). It is symbolically important that "mujahedeen" has come back into play. Formerly linked exclusively with the Afghan struggle against the Russians, it is being routinely applied in the Muslim world to the insurgency in Iraq.

I reminded him that in an October 2004 Internet statement Zarqawi had declared his allegiance to bin Laden and added that the two had agreed to join forces against "the enemies of Islam." Two months later, in an audiotape broadcast by Al Jazeera, bin Laden personally endorsed Zarqawi as his deputy and anointed him emir of Al Qaeda operations in Iraq. Al Qaeda has claimed responsibility for hundreds of deadly attacks against civilians and the military. Zarqawi's lieutenants have given interviews to Arab newspapers, announcing their aim to expel the Americans from Iraq and establish an Islamic state that would be a launchpad from which to overthrow impious Arab regimes. "We are fighting in Iraq but our sights are on other places, like Jerusalem," a Zarqawi operative told a reporter from *Al Hayat*. After interviewing scores of Zarqawi's lieutenants and close friends, his Jordanian biographer Fuad Hussein, who in the summer of 1996 spent time with him in Suwaqah Prison outside Amman, concluded that Zarqawi viewed Iraq as one pivotal battle in a bigger struggle to rid the Middle East of American presence and influence, reestablish the caliphate, and liberate occupied Muslim lands, particularly Palestine.

"Zarqawi plays a minor role in al-Muqawama," my Iraqi acquaintance insisted. "The Americans and their Iraqi cronies have inflated his importance because that serves their larger designs on Iraq and the Arab region. Many of the attacks attributed to Zarqawi's Al Qaeda are suspicious and carry the fingerprints of the enemies of Iraq." He claimed that American, British, and Israeli intelligence services were likely behind some of the suicide bombings against Iraqi civilians and the police; he reiterated his belief that the mujahedeen would only target the occupiers and their Iraqi collaborators.

Other Iraqis I met during the forum stressed that the American occupation of the country had militarized large segments of society, particularly the Sunni Arab community, which felt disfranchised and marginalized. The panel on Al-Muqawama al-Iraqiya presented documents written by Iraqi activists Muthana al-Darri and Salman al-Jumeili, both of whom have close connections to the resistance. They have concluded that al-Muqawama is predominantly made up of native-born mujahedeen; they based their findings on surveys of Sunni Arabs who lead the resistance and on interviews with families of the martyred mujahedeen.

As Kamal predicted immediately after the fall of Baghdad, the Iraq war seems to be giving rise to a new generation of jihadists. Kamal's point is being echoed by Western intelligence services. According to a 2005 report by the National Intelligence Council, the director of national intelligence's think tank, "[t]he al-Qa'ida membership that was distinguished by having trained in Afghanistan will gradually dissipate, to be replaced in part by the dispersion of the experienced survivors of the conflict in Iraq." This report took a year to prepare and includes the analysis of one thousand American and foreign specialists. It represents the consensus of the American intelligence community.

As it was for the Afghan Arabs, the raison d'être of the new generation of jihadists is armed resistance. Both differ from most militants in Kamal's generation, who put as much emphasis on doctrine as on action. The current Iraq fighters have already shown themselves to be bloodier than Afghan veterans, observing fewer limits on waging jihad. The increasing use of suicide bombings provides testament enough. Kamal believes that the militarization of jihadism has reached its climax in Iraq. "Look at what Zarqawi and his people are doing," he told me, "look at the beheadings and collateral killings of Iraqis who collaborate with the American occupiers." Particularly among young Muslims, American military occupation breeds more violence and extremism, which drive them into the arms of militants like Zarqawi.

Zarqawi and his lieutenants in Iraq recognize no distinction between civilian and military targets. They boast about killing Iraqi Shiites and Kurds, whom they accuse of collaborating with the American enemy. Islamists with access to Al Qaeda's inner circle acknowledge that initially even bin Laden was reluctant to agree to a merger between the Zarqawi group, called Jama'at al-Tawhid wa'al-Jihad, and Al Qaeda because it engaged in excessive sectarianism and bloodletting; Zarqawi excommunicated believers and justified collateral killing of Muslims "in order to ward off a greater evil, namely, the evil of suspending jihad," as he put it.

In contrast, bin Laden was not in favor of civil strife between Shiites and Sunnis, lest it distract from the confrontation against the Americans. As a militant Salafi, bin Laden undoubtedly harbors anti-Shiite prejudices, but he views Iraq as a pivotal front in his global jihad and has called on Muslim Iraqis and non-Iraqis of all ethnic and linguistic backgrounds to cooperate in opposing the pro-American order being installed in Baghdad. He has shown similar indifference to ethnic, sectarian, and ideological distinctions in issuing condemnations of Iraqis, including Sunni Arabs, who collaborate with the coalition forces. In December 2004 he released a videotape in which he declared, "The Iraqi who is waging jihad against the infidel Americans or Allawi's [former prime minister Ayad Allawi] renegade government is our brother and companion, even if he was of Persian, Kurdish, or Turkomen origin. The Iraqi who joins this renegade government to fight against the mujahedeen who resist occupation is considered a renegade and one of the infidels, even if he were an Arab from Rabi'ah or Mudar tribes."

Despite their earlier differences bin Laden and Zarqawi had found common ground by October 2004. According to firsthand accounts, as Zarqawi gained international notoriety and stardom among militant Islamists, he came under pressure by lieutenants and operatives to swear fealty to bin Laden and to merge with Al Qaeda: A formal merger with the parent organization would confer revolutionary legitimacy on Zarqawi and turn him from a mere field commander into a global jihadist on a par with Abdullah Azzam, bin Laden, and Zawahiri. Zarqawi's union with Al Qaeda could also bring a new crop of jihadist volunteers into Iraq, as well as funds. Even before the merger had been formally sealed, according to Zarqawi's Jordanian biographer, pro–Al Qaeda fighters, particularly from Saudi Arabia and Yemen, flowed to Iraq and joined Zarqawi's units, and wealthy Arabs contributed generously to the cause, which strengthened his network. More than anything else, this windfall motivated Zarqawi to pledge his full allegiance to bin Laden.

The windfall has also been bin Laden's. Before the invasion and occupation of Iraq, the global war had not been going well for bin Laden. He had trouble on two fronts: neither the ummah nor the international community supported his cause. Al Qaeda had also suffered crippling military blows and, with the 2001 fall of the Taliban regime in Afghanistan, lost its refuge and political patron. Along with senior aides, bin Laden was on the run. Most of his field lieutenants had been killed or captured, due in large part to assistance from the governments of Pakistan, Saudi Arabia, Yemen, and others, though the Western media have paid little attention to this fact. Equally important, Al Qaeda's centralized command-and-control decision making had been largely dismantled.

Iraq provided Al Qaeda with a new lease on life, a second generation of recruits, a way of extending its ideological outreach activities to Muslims worldwide. By appointing Zarqawi to the position of emir of Al Qaeda in Iraq, bin Laden could take credit for military successes there and rejuvenate his battered base. In the process he could reverse Muslims' hostile views of his jihad and further his goal of broadening his network's appeal to Arab and Muslim masses who feel enraged by the American occupation of Islamic territories. Internally besieged and in its final throes, Al Qaeda's new emphasis on Iraq is designed to tap into the reservoir of accumulated Muslim grievances against American policies. The same is true of Palestine, bin Laden's other newfound cause. After Israel assassinated Sheikh Ahmad Yassin, spiritual leader of militant Palestinian Hamas, on March 22, 2004, bin Laden, in a recording broadcast on Arab television networks, vowed revenge: "We vow before God to take revenge for him from America for this, God willing." U.S. policy ignores the "real problem," which is "the occupation of all of Palestine," he added as if America, not Israel, carried out the assassination.

Bin Laden has said that the Iraq war provided him with a golden and unique opportunity to defeat the United States and spread the conflict into neighboring Arab states like Syria, Jordan, Lebanon, and the Palestinian-Israeli front. He has portrayed the Iraqi resistance as the central battle in a Third World War, which the crusader-Zionist coalition started against the Muslim community: "The whole world is watching this war and the two adversaries: the ummah on the one hand, and the United States and its allies on the other." According to the May 2005 testimony by Seif al-Adl, Al Qaeda's number three and its overall military commander, who is under house arrest in Iran, Al Qaeda's leadership considers the Iraq war the most important development since September 11 in establishing the long-awaited Islamic state in the heart of the Muslim world. In a letter to Zarqawi that was intercepted and released by the U.S. authorities in October 2005, Zawahiri ranked the goals of jihad in Iraq in order of importance:

The First Stage: Expel the Americans from Iraq.

The Second Stage: Establish an Islamic authority or emirate, then develop it and support it until it achieves the level of caliphate—over as much territory as you can to spread its power in Iraq, i.e., in Sunni areas, in order to fill the void stemming from the departure of the Americans.

The Third Stage: Extend the jihad wave to the secular countries neighboring Iraq.

*The Fourth Stage: It may coincide with what came be-
fore; the clash with Israel, because Israel was established
only to challenge any new Islamic entity.*

But as Kamal tried to impress on me, the struggle in Iraq
is about much more than Al Qaeda. "It is about the future
of Iraq and the Arab and Muslim world," he told me. "Do
not reduce everything to the Al Qaeda organization. It is
al-shabab—Iraqis and other Muslims—who are fighting to
expel the occupiers from the heart of the Arab area. They
are not fighting to install bin Laden or Zarqawi as caliph." Al
Qaeda is a "minor faction that does not represent either Muslim public opinion or the dominant Islamist current. The
American global war on Islam has given Al Qaeda prominence which it otherwise would not have. Al Qaeda thrives
because of the new crusade. But once the war subsides, Al
Qaeda will be exposed for what it really is: very marginal."

Kamal continues to emphatically oppose some of the
methods practiced by Zarqawi's men—the kidnapping and
beheading of civilians, for example. These un-Islamic practices harm jihad's legitimate cause. "We insist that the resistance against occupation must remain civilized and respect
the rules of war," he said, his voice rising in agitation.

Kamal is not alone in his rejection of terror tactics in Iraq.
Other Arab and Muslim voices can increasingly be heard
condemning Zarqawi and appealing to Iraqis and Arabs to
oppose his murderous ideology. Tantawi, the aforemen-
tioned Grand Imam of Al-Azhar in Cairo, called on the international community to put an end to terrorism in Iraq

and to punish Zarqawi and his men for killing civilians, which violates Islamic precepts. Tantawi insisted that those members of Al Qaeda convicted of murder must be vigorously pursued, captured, and sentenced to death. In two separate statements the imprisoned leaders Tanzim al-Jihad and al-Jama'a al-Islamiya have denounced Zarqawi for killing civilians, including diplomats and government employees, and have accused his organization of trying to annihilate the Shiites rather than liberate Iraq.

After three of Zarqawi's suicide bombers struck three hotels in the normally tranquil capital of Jordan, Amman, on November 9, 2005, killing fifty-seven people and wounding one hundred or more others, thousands of Jordanians rallied in Amman and other cities, shouting, "Burn in hell, Abu Musab Zarqawi." Other protesters cruised Amman's streets, honking their horns and chanting, "Death to Zarqawi, the villain and traitor!" and antiterrorism slogans. Some of the protesters in Amman carried placards reading simply WHY? Why would Zarqawi target their country, where so many people had supported his jihad in Iraq?

More telling were the reactions of residents of Zarqawi's run-down industrial hometown, Zarqa, who expressed anger that one of their own could massacre civilians—Arabs and Muslims—whose only sin was being in the wrong place at the wrong time. The wife of Zarqawi's brother told an *Al Hayat* reporter that she was dismayed by the carnage she had seen on television. Um Arwa, a neighbor of Zarqawi's household said, "What happened has nothing to do with Islam." Another neighbor named Raouf Ahmad Omar, who has lived in Zarqawi's neighborhood for twenty-five years cursed his old acquaintance. "Those do not fear Allah...From now on the mosque preacher must speak out and condemn terrorism. There is no war for them to carry out such criminal operations." A middle-aged man named Zuheir Najjar told another journalist that the Amman attacks backfired on Zarqawi: "What does an attack on a wedding with women and children have to do with fighting the Americans?" Other residents regarded the suicide bombings as barbaric and un-Islamic. "Any person who would do such an act must be considered a heretic," said Abu Ibrahim, a fifty-six-year-old merchant standing outside his shop, several hundreds yards from Zarqawi's high-walled house, where his relatives still live. Zarqawi has even been disowned by members of his family, part of the influential Bani Hassan tribe. "We sever links with him until doomsday," wrote one in a Jordanian newspaper. According to tribal traditions, some family members may now seek to kill him.

Initially, Zarqawi's organization said the bombings put the United States on notice that the "backyard camp for the crusader army is now in the range of fire of the holy warriors." Feeling the heat of public opinion throughout the Arab world, Zarqawi's group took the rare step of trying "to explain for Muslims part of the reason the holy warriors targeted these dens"—meaning the three hotels. "Let all know that we have struck only after becoming confident that they are centers for launching war on Islam and supporting crusaders' presence in Iraq and the Arab Peninisula and the presence of the Jews on the land of Palestine," Al Qaeda in

Iraq said in an Internet statement. In a subsequent Internet statement, Zarqawi emphasized that his organization was not targeting fellow Muslims. "We did not and will not think for one moment to target them," he said. But the statement showed how tone-deaf to public opinion Zarqawi is, for it also promised more "catastrophic" attacks. The killing of civilians in Jordan triggered an unprecedented torrent of angry and emotional responses in the Arab world. In a moving article entitled "I Am Also Zarqawi," published in the pan-Arab nationalist newspaper *Al-Quds al-Arabi*, Amjad Nasser, an artist born and raised in Zarqa, paints a nostalgic portrait of the town. "I am also Zarqawi, like many other ordinary people in Jordan, but we are made of a different fiber than the one who hijacked the name of the city and turned it into a banner of blood and death," he wrote.

There are real indicators that Arab and Muslim public opinion has turned again against Zarqawi and Al Qaeda. The bombings and killing of civilians in Saudi Arabia, London, Indonesia, Egypt, and now Jordan have all triggered a torrent of angry responses by ordinary Arabs and Muslims across the political and religious spectrum. In a survey of more than 1,000 Jordanians conducted for the newspaper *Al-Ghad*, more than 87 percent of the respondents said they now considered Al Qaeda a terrorist organization. (In previous surveys in Jordan, Al Qaeda had enjoyed approval ratings of 60 percent and higher.) Other polls in various Arab countries confirm this change of opinion.

Even Abu Mohammed al-Maqdisi, Zarqawi's spiritual and ideological mentor, has publicly opposed his disciple's use of terrorism against civilians. Al-Maqdisi is considered the mufti of the Jihadi-Salafi current that has inspired a host of jihadists, including Zarqawi. He spent three years in prison with Zarqawi—1996 to 1999—during which he shared his jihadist writings with the younger man and had a seminal influence on his ideology. In interviews with Arabic-language newspapers and Al Jazeera television network, al-Maqdisi said that violence that did not differentiate between women and children, civilians, soldiers, and American troops, was wrong. "The kidnapping and murder of relief workers and neutral journalists has distorted the image of jihad. They make the mujahedeen look like murderers who spill blood blindly."

In two recently released statements from his prison cell in Jordan, al-Maqdisi struggles to strike a balance between criticism of Zarqawi's mass killing of Muslims and praise for his resistance against the Americans. He warns his former pupil against extremist tactics that alienate both friend and foe and play into the hands of the enemy. "The Shiites and Kurds," al-Maqdisi advises Zarqawi, "are not the enemy; the American occupiers are. Do not lose sight of the nature and character of the struggle in Iraq, because Iraqis know what is best for their country; know your place and avoid leveling threats against other nations and people lest you turn Iraqis and the whole world against you and your mujahedeen." Explicit in al-Maqdisi's fatherly advice to Zarqawi is the urgent need for the latter to refrain from mass killing of Muslims and to reassess his "indiscriminate" attacks that distort the true jihad. Time and again he cautions his pupil against relying on unsound advice. This public criticism of Zarqawi by one of his foremost allies has reportedly caused considerable disquiet among Zarqawi's close lieutenants and once again reveals fissures within the jihadist movement.

But the most telling rebuke was revealed in a communication intercepted by American authorities. In July 2005, Zawahiri dispatched a 6,000-word letter to Zarqawi, chiding him for alienating Arabs. "In the absence of this popular support," Zawahiri wrote, "the jihadist movement would be crushed in the shadows." Anyone who doubts the authenticity of Zawahiri's letter should read *Knights Under the Prophet's Banner,* his memoir, in which he calls upon militants to fully integrate into society and lead the ummah.

Not only have militants of the Al Qaeda variety joined the fight in Iraq. The occupation has been effective in recruiting others who oppose America's policies; the war has radicalized public opinion throughout the Arab and Muslim world. Young Muslims unconnected to Al Qaeda but enraged by the U.S. military presence in the heart of Islam have become receptive to the militants' call. What the American occupation of Iraq has done, Kamal argues strenuously, is drive young Muslims who had nothing to do with Al Qaeda to take arms against America. Kamal drew my attention to what he termed the "social Islamic network"— Muslims who reside far away from one another, yet share common ties and a moral obligation to defend the ummah when it is threatened. "There are many independent networks of Muslim activists in various countries without formal affiliation with Al Qaeda that could carry out attacks against American interests all over the world," Kamal told me in November 2005. "These decentralized networks are united by opposition and hatred of U.S. policies and can be activated anytime, anywhere, without prior warning, like in Indonesia, Madrid, and London. The new crusade has triggered a new type of war without borders. The Al Qaeda organization supplies the ideology but al-shabab act on their own with no centralized supervision, or control." Kamal told me that those independent networks do not know one another and do not coordinate their attacks.

As I was completing this book, I called Kamal and questioned him further about this social Islamic network. How do these young Muslims get drawn into jihadist causes? Kamal replied by giving me a lecture on the "charisma of thought." Islamic fiqh stipulates that when aggressors invade the territories of the ummah, Muslims must wage jihad. In this, jihad becomes a permanent and personal obligation, or fard'ayn, Kamal stressed; there is no way out of this obligation. Imagine then when the military aggression is coupled with a cultural threat: "This situation creates terror among young men who embrace the charisma of thought that calls for defiance and resistance."

In late 2004 I had traveled to Majdal Anjar, a town of twenty thousand located in Lebanon's Bekaa valley, forty miles east of Beirut. Situated near the Lebanese-Syrian border, the town had gained national notoriety a few months before because of clashes between activists and the Lebanese authorities; the police had arrested several residents, accusing them of forming an Al Qaeda network and recruiting young men to wage jihad in Iraq. I had expected the villagers to be guarded about expressing their real feelings about the events in Iraq. Instead, the young men I encountered talked freely about jihad in Iraq as being a religious duty and an honor, and about how they would welcome martyrdom. They were angry because the Lebanese government had cracked down on men wishing to travel to Iraq; they suspected their politicians had been bought by the Americans.

The young men were particularly outraged by what they saw as the massacre at Falluja. That November, between ten and fifteen thousand American soldiers and marines, backed by newly trained Iraqi forces, invaded the city and fought pitched battles with Iraqi insurgents. Although the Americans took control of Falluja, the city was partially destroyed and thousands of Iraqis were killed. Many were shocked by the extent of the destruction and by the huge number of deaths in Falluja, a city famous for its mosques and minarets. Like others, the young men of Majdal Anjar had watched the tragedy unfold on their television screens. Thanks to coverage brought by Al Jazeera, they were able to follow the battles in Falluja block by block and street by street.

They would do anything, they declared, to fight alongside their Iraqi counterparts. Some of them confided that if they could afford the taxi fare from Majdal Anjar to the Syrian-Iraqi border, they would leave without even telling their families. They all expressed admiration for the few mujahedeen from their town who had been martyred in Iraq.

Based on these conversations, as well as numerous other

talks with Arab men and activists, I am convinced that were it not for logistical and technical challenges, more volunteers from the Middle East and Europe would travel to Iraq than ever went into occupied Afghanistan during the 1980s. Pressured by the U.S. government to prevent foreign fighters from crossing its border into Iraq, the Syrian regime has said it has apprehended more than eight thousand men from the Middle East and Europe and repatriated them to their countries of origin. Among them were four thousand Syrians, fourteen hundred Arab foreigners, and twenty-five hundred Iraqis. American officials remain unconvinced that Damascus has fully secured its border. Other recruits are traveling to Turkey and Iran and crossing into Iraq, often through the large unguarded portions along Iraq's vast border. While the majority are from countries in the Persian Gulf, mainly Saudi Arabia and Yemen, many are from North African countries, Syria, Lebanon, Egypt, the Sudan, and Jordan. Moreover, scores of young Muslims from European countries, principally France and Britain, have already fought in Iraq, with a large pool of potential recruits searching for ways to get there.

A month after the recapture of Falluja, Kamal gave vent to his rage in a commentary entitled "The Steadfastness of Falluja and Rewriting of History," which appeared on the Center for Islamic Studies Web site: "What happened in Falluja was a struggle between eternal Truth and evil which has no future," Kamal wrote angrily. Falluja taught "American arrogance" an important lesson about willpower and resistance, he added, and "exposed America's ugly racism and hatred of people."

I had followed Kamal's writings closely over the years; this was one of the most vitriolic articles I had read. He was outraged by an image of an American soldier killing an injured Iraqi fighter in a mosque. "What is happening in Iraq is collective genocide against the Sunni Arab community. The American model of democracy will be implemented with blood, massacres, and use of weapons banned internationally." America's intention to plant the seeds of democracy in Iraq is big lie, at least according to Kamal. He told me evidence exists that "the Americans came to Iraq with a sectarian secular program designed to build a divided and weak sectarian society and rely on the Shiites and Sunni Kurds to do their bidding." For him, whatever the actual statistics and real numbers, Sunni Arabs represent a majority of Iraq's population of twenty-three million and represent the only hope for a unified Iraq, one with an Arab-Islamic identity. There is no doubt, Kamal assures his readers, that armed Sunni resistance will defeat the American occupiers and their Shiite and Kurdish clients.

Once again I called Kamal and tried to provoke him. "You say the Iraqi resistance will prevail, while the Americans say they are winning," I told him. "America cannot win in Iraq," he replied without reflection. "It does not possess knowledge of the Arab mood and social structure. America does not understand the psychology of the Arab and Muslim people. American strategists err dangerously by comparing us to the Japanese and German people after World War II. We cannot be molded and subjugated by outsiders. When we are attacked, we rise up to the challenge."

In the wake of the July 2005 London bombings, the international media scoured the neighborhoods where the four British-born bombers had lived, searching for clues that would explain the attack. The American-British occupation of Iraq and suffering of Iraqis figured prominently in the answers given by some of the bombers' acquaintances. The *New York Times* quoted Sanjay Dutt, twenty-two, who was grappling with why his friend Kakey, better known to the world as Shehzad Tanweer, had decided to kill his countrymen. "He was sick of it all, all the injustice and the way the world is going about it. Why, for example, don't they ever take a moment of silence for all the Iraqi kids who die?" "It's a double standard, that is why," answered a friend.

"We've got to look at the reasoning behind these things," said Saraj Qazai, a twenty-five-year-old owner of a boutique. "There's no denying it's payback for what's happened in Iraq and Afghanistan. You've been bombing people for the last two to four years, so you are going to get a backlash. England is a great country and we love it to bits but do we love this government? No. There were twenty-four Muslims killed in Iraq today; there will be more tonight and more tomorrow."

Of course, it would be reductive and simplistic to claim that the London bombings were a result of the rage generated within the global Muslim community by the Iraq war. But the comments from these young people throw light on the mood of young foreign-born Muslims. European officials and observers have already voiced their concern that Iraq is becoming a recruiting tool, attracting hundreds of second-

and third-generation European Muslims. At the same time, a July 2005 report from Britain's conservative Royal Institute of International Affairs concluded that backing the United States in the war in Iraq had put Britain at greater risk from terror attacks. The British government understandably rejected the report's conclusions. But according to the institute, a majority of Britons believe that their country has suffered by playing "pillion passenger" to the United States.

As in Europe, the upper echelons of the American government seem oblivious to the serious repercussions of the Iraq war on American security. American, European, and Arab intelligence services agree that Iraq has replaced Afghanistan as the training ground for a new generation of "professionalized" jihadists who are sharpening their technical skills. A June 2005 classified assessment by the CIA states that while Iraq has in many ways assumed the role played by Afghanistan in the 1980s and 1990s, attracting militants from Saudi Arabia and other Muslim countries, it may ultimately prove an even more effective training ground, because it acts as a real-world laboratory for urban combat. The consensus among intelligence agencies is that Iraq is gradually replacing other fronts as a forward base for the new global jihad and as a gathering point for a large concentration of active jihadists. "Our policies in the Middle East fuel Islamic resentment," U.S. Vice Admiral Lowell E. Jacoby, director of the Defense Intelligence Agency, told the Senate Select Committee on Intelligence in 2005. The American government has launched a high-level internal review to assess the effect of what one government official called "the bleed-out" of hundreds or thousands of Iraq-trained jihad-ists back to their home countries throughout the Middle East and Western Europe. "It's a new piece of a new equation," a former senior U.S. official acknowledged.

⚓

The equation may not be so new—it may be as ancient as the roots of jihad—or it may be the place where all the equa-tions, new and old, coincide. I began this book by setting out to find Kamal, whose journey, I felt, epitomized the recent history of jihadism. After attempting to bring about change by violent means, he had laid down his arms and struggled for a way to participate in the political process at home. He had played a pivotal role in ending the war between Egyp-tian militants and the authorities. Kamal had believed that Islamists needed to change course, to travel a dramatically different road, to make a clean break with their violent past. Until the end of 2001, he had stressed the need for dialogue with the Western powers, including the United States, and criticized the Afghan generation for its excessive militarism and lack of political vision.

Now Kamal has become convinced that the outcome of the clash between the Muslim community and the American-led coalition will shape the identity and future of Arabs and Muslims for years to come. It is an epic struggle, one that must be joined at all costs. "Is this clash really inevitable?" I pressed him. "As long as America occupies Muslim territo-ries, there will be conflict," he retorted. "We—Muslims—

have an obligation to support al-Muqawama. I support the expulsion of America from Iraq."

Before 2001, Kamal would have cheered any power, including the United States, that turned its guns against local Muslim rulers—the near enemy. With the U.S.-led war on terror, which encompasses in the occupation of Baghdad, the former seat of the Islamic caliphate, "it is no longer acceptable or logical," he says, "to contribute to what appears to be part of an American plot to spread instability in the Arab and Muslim regime."

To Kamal and others, the struggle in Iraq has little to do with Al Qaeda or with any specific group. It symbolizes generalized American aggression against a Muslim state. He is convinced that resistance to "atrocities" is the only way forward. I asked Kamal if his call for jihad against the Americans made sense in light of the imbalance of power between Muslims and the greatest military force in history. From the time I met him in 1999, Kamal had criticized his fellow jihadists for their impulsiveness, their unwillingness to distinguish between means and ends. He had always urged Islamists to employ strategic thinking. After his release from prison he had come to the conclusion that politics and social action could be as effective, if not more so, as armed struggle. I would often hear him recite the famous saying by that nineteenth-century Prussian military officer Carl von Clausewitz, that war was a continuation of politics by other means— whether used for defensive or offensive purposes.

Now, six years after our initial conversations, he was

conceding that the United States was militarily more powerful and the fight unfair. "Our power is not comparable with the enemy's. I do not advocate a hasty or rushed confrontation with America. But this is war. The new crusade put us in a corner and left us no choice but to defend our values and religious icons. War is forced on us. What do you expect us to do in the face of aggression against the ummah?" he asks, his voice filled with anger, pain, and frustration. "I want you to know that I have no operational links with any paramilitary organization. I am talking as an Islamist observer who has his finger on the pulse of public sentiments. Al-shabab are dying to join the fray. They want to defend their injured ummah. They cannot remain on the sidelines, while the crusaders defile their lands and homes and honor."

Kamal's views have come full circle. As he did a generation ago, he believes that armed confrontation is inevitable—that jihad is the only journey to repulse external aggression against the ummah. Why? I ask. He told me that in the fiqh, *hizb al-din*, "the party of God," comes before life itself: "If and when our religion and the ummah are threatened, we sacrifice life to defend them. Everything pales by comparison with our Islamic doctrine and nation."

Kamal's jihadism nonetheless remains more limited in scope and character than that of Al Qaeda jihadists. He called for the liberation of Iraq, not an open-ended confrontation with Western powers. "I feel sorry for the U.S. soldiers who are being killed on daily basis in Iraq," he said once out of the blue. "They are victims of the imperial policies of the new conservatives." And he urged me to deliver the message to the American people that what Muslims and Americans have in common outweigh any differences and conflicts of interests. He said he hoped and prayed that the American people would reclaim their values of tolerance and respect for the rule of law, and that they would put an end to the new crusade against Muslims. "There is a way out of the current deadly embrace between the United States and the Muslim world. We must acknowledge the common ties that bind us and our shared humanity." Later, he added, "We have a unique culture, a unique religion, a unique history, and a unique civilization. The Western nations can learn from us if they open their hearts and minds."

"What about your own journey?" I asked him. "At a historical moment my generation thought that we could transform our society by force," was his reply. "We were young and full of zeal and fervor. But we learned the hard way that our vision was narrow and limited. Our imagination and enthusiasm got the best of us. Now we have moved from the narrow to the real world. We have learned the art of compromise. We cannot change our reality without changing ourselves."

I asked him what he says when those in the West argue that all Islamists and jihadists are inherently violent evildoers? "I want you to know that violence was not part of my character and my generation's, contrary to what you read and hear about us. We were not born with a violent gene. We were not inherently violent. Our violence was a product of a political vision, specific conditions, and circumstances.

"The jihadist is not a beast or a new barbarian. He goes on that journey not because he wants to but because he must. The Qur'an tells us that fighting is repugnant. The journey is paved with hardships—painful and costly. Please do not buy into the notion perpetrated about us in the West that we worship death. We possess no cult of death. We love life. We love our families. But we want to live with dignity. We are willing to sacrifice our life because we want to live as free men rather than being enslaved. The jihadist who makes the supreme sacrifice to defend his religion and home is an exceptional human being—full of humanity."

EPILOGUE

ITINERARY of a JIHADIST

1987. A former Lebanese army officer named Mustapha Darwish Ramadan immigrates to Denmark for economic opportunities. Beirut-born and of Kurdish descent, Ramadan is married to a woman from Majdal Anjar, a town of 20,000 located in Lebanon's Bekaa Valley near the Syrian border. His marriage strengthens his connection to Majdal Anjar, which is a bastion of Islamic fundamentalism and extremist Jihadi-Salafi thought.

Ramadan spends fourteen years in Denmark, developing contacts with radical Islamists across Europe. Eventually he joins a chapter of Ansar al-Islam, a small Kurdish-Sunni organization with roots in northern Iraq. During its early days Ansar al-Islam had been involved with local conflicts only, but by the time Ramadan joins it, it has recast itself as an international force in Islamic militancy and expanded into Europe.

1997. Thieves steal more than $300,000 from an armored car in Copenhagen. Investigators discover that Ramadan is one of the culprits. He is arrested shortly before attempting to flee Copenhagen on a flight to Amman, Jordan, and convicted of robbery. He serves three and a half years in prison.

June 2001. Ramadan is released from prison. He immediately becomes involved in another heist, robbing a money-transfer store of about $15,000. This time, he escapes arrest and leaves Demark.

Summer 2001 to early 2003. Ramadan's whereabouts are unknown. He is believed to be either in Jordan or Lebanon.

Winter 2003. Ramadan turns up in Majdal Anjar, whose townspeople are struck by dramatic changes in his appearance and conduct. He is abstemious, dresses like an Afghan mujahid, and has let his beard grow. He immediately begins proselytizing among young Salafis, winning them over to his call for *takfeeri*, or excommunication of anyone opposed to the cause of jihadism.

April 9, 2003. Baghdad falls to American-led coalition forces.

June 2003. Differences within the community force Ramadan and his family to leave Majdal Anjar for the nearby village of Kafir Zaid. By now he has attracted a core of young loyalists who are prepared to go to Iraq and fight against its occupiers.

According to firsthand accounts, when a volunteer decides to go to Iraq, he first goes to a mosque to initiate contact with those who can help him make his way into Syria. Once in Syria, jihad volunteers generally spend two or three weeks in a safe house before being sent into Iraq. Some are given a crash course in the operation of Kalashnikov rifles, rocket-propelled grenade launchers, and remote detonators; the training takes place in camps located in the Syrian Desert, near the Iraqi border. Others receive no training and will drive and detonate cars rigged with explosives.

July 2003. A Yemeni named Abu Mu'az travels from Syria to Kafir Zaid and meets with Ramadan. Abu Mu'az coordinates the transport of young men from the Majdal Anjar region through Syria to Iraq, which shares a long, porous border with Syria. The Sunni tribes with branches on either side of the Syrian-Iraqi border view that border as little more than an inconvenience; they have long made a living smuggling goods, services, and people. This route becomes the preferred route into Iraq for jihadists, who are guided down the Euphrates river valley from the border town of al-Qaim, through the towns and cities of Haditha, Hit, Ramadi, and Falluja, and eventually arrive in Baghdad. (Others cross into Iraq through the northern town of Tal Afar and the capital of the Northern Province, Mosul.)

August 2003. Ramadan enters Iraq through the Euphrates river valley. He is already well known for his work for Ansar al-Islam during his years in Denmark and for his recruitment work in Majdal Anjar. Therefore, unlike other foreign (non-Iraqi) volunteers, Ramadan is immediately given security clearance. He rejoins Ansar al-Islam, whose surviving members have moved to the Anbar Province in western Iraq, the heart of Sunni resistance. His son, Mohammed, fifteen years old, accompanies him.

Fall 2004. According to Iraqi-Kurdish sources and European intelligence officials, Ramadan has emerged as a senior leader of Ansar al-Islam, which Americans and their Iraqi allies suspect of having carried out more than forty suicide bombings and other attacks, resulting in more than one thousand fatalities. Ramadan now operates under the name Abu Mohammed al-Lubnani, or "father of Mohammed the Lebanese," and has become a close aide to Abu Musab Zarqawi, emir of Al Qaeda in Iraq.

Al-Lubnani continues to recruit volunteers from Majdal Anjar, smuggling them into Iraq. Several of these recruits, among them Hassan Sawan, Ali al-Khatib, and Mohammed Nouh, are killed in combat and celebrated as martyrs in Majdal Anjar. Al-Lubnani is now being actively hunted by American and Iraqi forces.

Early 2005. A Web site run by militant Islamists reports that al-Lubnani has been killed. The claim is greeted skeptically by American officials, who believe it may be disinformation. However, credible Iraqi-Kurdish sources confirm that al-Lubnani perished in an American airstrike. A few months earlier, his son Mohammed met a similar fate.

Note: Dates in this itinerary are approximate and based on the sources indicated in the notes to this book.

Foreign jihadists, such as Mustapha Darwish Ramadan, alias Abu Mohammed al-Lubnani, make up a small portion of the insurgent forces in Iraq—less than 10 percent; the rest are native Iraqis. Nonetheless, the vast majority of suicide bombers in Iraq are thought to be foreign jihadists, principally from Saudi Arabia and the Gulf States, Syria, Jordan, Lebanon, and North Africa. A number of these jihadists have emigrated from European countries. Since the handover of sovereignty in Iraq in June 2004, there have been more than 500 car-bomb attacks, killing and injuring about 10,000 people.

AFTERWORD

FIVE YEARS AFTER September 11, 2001, Egypt was my third stop on a fifteen-month research journey through the Middle East. My arrival in Cairo coincided with a torrent of protests in Egypt and the rest of the Muslim world over remarks made by Pope Benedict XVI that implicitly linked Islam and violence, particularly in regard to jihad, or "holy war." In a visit to his native Bavaria in southern Germany, the Roman Catholic leader had quoted a fourteenth-century Byzantine Christian emperor, concerning the founder of the Muslim faith: "Show me just what Mohammed brought that was new, and there you will find things only evil and inhuman, such as his command to spread by the sword the faith he preached."

I called Kamal to find out more about the different ways in which the pontiff's words were heard inside mosques, alleyways, cafés, classrooms, and underground bunkers. I knew he would have a good understanding of Muslim public opinion, particularly that of militant Islamists. To my surprise, he invited me to attend the Friday prayers with him at Al-Azhar Mosque in the heart of old Cairo, though he knows I am not a Muslim. He warned me that there would be protests after the sermon, and that they could turn violent.

When Kamal and I arrived at Al-Azhar—one of the oldest surviving Fatimid architectural landmarks (completed in A.D. 972) and most respected Islamic institutions of higher learning—hundreds of uniformed and civilian security officers had cordoned off the area. Antiriot police stood ready by dozens of military buses and vans. I hesitated before slipping through the well-guarded entrance gate built in the eighteenth century, where students were once shaved.

I deposited my shoes at a makeshift stall managed by an elderly man, and then Kamal and I entered a large courtyard surrounded with porticos supported by over three hundred columns, their colored marble intricately carved and seamlessly fitted into stunning patterns. Surely Al-Azhar ("the splendid") lives up to its name architecturally, though it has stagnated intellectually in the last few hundred years, its scholars shying away from any critical effort at religious renewal and reformation. We tiptoed our way inside the prayer hall and to the front, next to the imam of Al-Azhar Mosque, Sheikh Salah al-Din Nassar.

Dressed in Al-Azhar's attire (a blue robe and a red-and-white turban), Sheikh Nassar delivered a sermon about the urgent need for Muslim unity. "Unity equals power," he said in a high-pitched voice. "If Muslims close ranks and unite, no one in the world would dare to attack them and insult their religion and Prophet. Those who accuse Islam of intolerance and violence are either ignorant or full of enmity.

"No, Islam was not spread by the sword," he said, gently touching the microphone. "One of the fundamental tenets of Islam is that there is no compulsion in religion—you have your religion and I have mine." Believers nodded their heads.

As soon as Sheikh Nassar concluded his sermon with "Peace be upon you," roars of "Down with the pope, down with the Vatican!" echoed from one end of the mosque to the other. Hundreds of protesters rushed outside with placards proclaiming that the Vatican's war on Islam was an extension of Bush's war on Islam. People chanted, "Where are you, Muslims? The pope is waging a new crusade against Islam." I watched as Islamic activists, blocked by security personnel from spilling out onto the streets, went around and around inside the gate, carrying a young man and chanting after him, "Oh Mubarak, where are you? Where are you? Mohammed's religion is your religion too."

Radical Islamists, Muslim Brothers, and independents were at least as angry at Arab rulers as they were at the pope. They accused Arab governments of being pliant and submissive and of not defending the faith and their nations. Iraq, Palestine, and Lebanon were cited as examples of collusion between Arab rulers and their Western "masters." When I asked a young teenager named Hussein, who had dark curly hair and piercing dark eyes, why he was protesting, he said, "I feel aggrieved because our prophet is being smeared, and none of our leaders are defending him." Hussein claimed he did not belong to any political organization; "I am just a believer," he said.

Speakers reminded their listeners that they must avoid violence. "Our resistance must be peaceful," barked Magdi Hussein, a leader of an outlawed radical Islamic group, while gripping the microphone tightly. It was a relief to see radical Islamists showing a restraint and political maturity they had lacked before, but I asked Kamal—who was participating in the protests—whether he expected there to be a violent response to the pontiff's remarks. He said that Christian Arabs might be targeted by outraged Muslims who lumped them together with the Roman Catholic Church. The larger danger, however, was that young Muslims would be radicalized by accumulated grievances and would join militant groups such as Al Qaeda. Kamal pointed out that Muslims could not view the pontiff's statement in isolation. Coming on the heels of President Bush's declaration of a "war of civilization" on "Islamic fascists," the pope's comments seemed to supply religious justification for a Western onslaught against Islam and Muslims.

Amidst the protests, I snuck away to visit with Sheikh Nassar and other Al-Azhar clerics in Nassar's private office, which was filled with classical Islamic texts and decorated with beautifully lettered Qur'anic verses. A distinguished man in his seventies, with dark features, Sheikh Nassar was employed by the Egyptian government and therefore had to be careful in his sermon. "Enough is enough," he said, tapping his fingers on the table. "The pope talks about tolerance. Does insulting Islam and its prophet reflect any religious tolerance? We—Muslims—respect and recognize the people of the book [Jews and Christians] and their prophets."

Afterward, Kamal and I took a long stroll down the narrow alleyways of old Cairo. "The pope's diatribe is very dangerous because it plays with fire—with religion," Kamal volunteered. "A clash of religions is the most perilous and most difficult to extinguish. Everyone will lose"—a fascinating response from a militant Islamist who once believed in the clash of civilizations himself.

Coming face-to-face with the perils of daily Egyptian life—the widespread poverty, the sewage, the unpaved streets, the architectural gems tottering on the verge of collapse, the pollution, the congestion, and the deadly traffic—it was painfully obvious that impoverished Egyptians and other Muslims fall back on religion as a refuge from and shield against their unforgiving reality. It was also painfully obvious that radical Islamists like Kamal exploit perceived Western hostility toward Islam in order to recruit Muslims to their cause. For example, although Pope Benedict XVI expressed regret for any hurt his comments caused, Islamists demanded a more explicit apology and called for further protests. Responding to the Egyptian-born and Qatar-based Islamic scholar Yusuf al-Qaradawi's call for a "Day of Anger," thousands turned out at Al-Azhar Mosque the next Friday.

Kamal, along with two of his friends, escorted me to Al-Azhar. The Muslim Brotherhood, one of the most powerful Islamic organizations in the Arab world, had provided scores of young Muslims with headbands that read WE ARE ISLAM'S SOLDIERS, and hung the mosque courtyard with banners proclaiming STUDENTS OF THE MUSLIM BROTHERHOOD DENOUNCE THE POPE'S REMARKS.

The imam at Al-Azhar reminded the believers that Ramadan, Islam's holiest month, a time of fasting and abstinence, had begun that week, and that they should repent and purify their hearts and souls. He concluded his sermon by saying, "We will not accept the pope's apology because it is not enough. He must erase the quote that linked Islam with violence so that future generations will not use it." Instantly, the prayer hall was transformed into a political rally. The crowd roared with a deafening chant: "With our blood, with our soul, we sacrifice for Islam."

The first speaker, Mustafa Bakri, a nationalist journalist and parliamentarian, asserted that the West is waging a "crusade" against Islam. "The highest authority in the Catholic Church is supplying religious justification for Bush's war. The other is the killer. The other is the aggressor. We [Muslims] must unite. We must do away with compromise and reconciliation with the other," shouted Bakri, the only non-Islamist speaker present. Next, Mohammed al-Beltagui, a Muslim Brotherhood parliamentarian, dressed in a suit and tie, reminded the crowd that the pontiff's remarks were not an innocent slip of the tongue. "We must stand up and support resistance movements in Palestine, Lebanon, and

Iraq. If we unify our ranks, the ummah, or global Muslim community, would succeed in repelling the aggressors who are robbing its treasures. We will not accept the pope's apology...He must concede that he has blundered," al-Beltagui reiterated.

The crowd was then brought to tears by a historian and Muslim Brotherhood speaker, Sheikh Gamal Abdul-Hadi. A short man in his late sixties, bearded and dressed in traditional Islamic attire, Sheikh Abdul-Hadi cried, "Where is the ummah? Where are Muslim leaders, Muslim clerics? They must sever diplomatic relations with the Vatican, with the Americans, and with the Germans. They must call for jihad to fight. What are you waiting for?" I noticed several men dressed in white, hooded lest they be identified by the local authorities, with the same statement written across their chests: WE ARE PREPARED TO BE MARTYRED FOR ISLAM.

Known for its organizational skills, the Muslim Brotherhood had left nothing to chance at the rally. They kept a tight balance between students who chanted anti-pope slogans and speakers who incited the crowd. Everyone acted in unison. As part of their recent efforts to demonstrate their evolution from a patriarchal structure into a politically and socially inclusive organization, the Brotherhood also mobilized an impressive contingent of female supporters. Veiled and vocal, the women added color and texture to the protests.

Finally, a speaker representing the Islamic International Federation of Non-Governmental Organizations called on Turkey to refuse the pope's planned visit to the country in late November. And Omar Azzam, a leader of the banned Islamic Labor Party, said, "We must thank the pope for what he said because he awakened the ummah." These were the only representatives of other political parties to participate in the Brotherhood's rally.

Afterward, at a coffee shop nearby, Kamal said he was disappointed by the low turnout; he had expected tens of thousands of protesters. "Aren't you exploiting the pope's remarks to garner political gains at home?" I asked him. "So what?" he retorted, sipping a glass of Egyptian tea. "The pontiff has given us an opportunity to mobilize the ummah."

"Obviously you are staging the protests for media consumption," I said. "Where are the crowds of angry Muslims? They are nowhere to be seen—just small groups of politicized Islamists and disgruntled activists."

"No; I hope you do not misunderstand me," Kamal said. "We are not inciting people. Ask anyone here in old Cairo and they will tell you that they are hurt." Kamal grabbed two Egyptians who happened to be walking by, and put the question to them. Walid and Abdul Nasser said they felt offended by and angry at the pope's remarks about Mohammed.

"You see?" Kamal said with a triumphant smile. I could tell he didn't really understand what I was getting at.

"But your inflammatory rhetoric harms interfaith relations between Muslims and Christians in the long term," I said.

"Well, the pope's remarks have already done irreparable damage. We [Islamists] have a responsibility to our audience,

which expects us to defend the faith. We cannot think rationally while our prophet and religion are being smeared. The
pope's anti-Muslim statement has left us with no choice but
to protest."

"Well, where do we go from here?" I asked Kamal and
Hazem Salem, a human rights advocate in his thirties who
had accompanied us to the coffee shop.

"If the pope makes an unambiguous apology, if he can-
cels the hurtful quotation, that would resonate positively
with Muslims," Kamal volunteered.

"Some broken things can never be repaired," retorted
Hazem, a progressive leftist in the opposite camp of Kamal.
"The problem is that Muslims are taking a look at Pope
Benedict XVI's past and do not like what they see. The crisis
has widened the cleavage that exists between the world of
Islam and the Christian West. Exacerbating the crisis is that
Arabic Islam and non-Arabic Islam in Iran, Pakistan, In-
donesia, and Malaysia outbid one another to show that they
are the real defenders of religion. The result is that the rever-
beration of the pope's words echo near and far."

Kamal nodded.

Indeed, Kamal's earlier warnings have proven accurate.
In the fall of 2006 I interviewed scores of Muslim activists,
human rights advocates, mainstream and militant Islamists,
liberals, and ordinary citizens. Most told me that the West,
particularly the United States, is waging a modern crusade
against Islam. From high-school teachers to taxi drivers, ordinary Muslims saw America as a new colonial power. Few
Muslims accepted the American narrative that touts democ-

racy and freedom. They viewed America's military presence in Iraq, the Arab heartland, as a sinister plot to divide the world of Islam and subjugate Muslims.

"Look at what America is doing in Iraq," said Hazem excitedly. "America is using democracy as a mask to colonize Muslim lands and to steal our oil." I reminded him that President George W. Bush claims he is promoting democracy in the Arab world. "No, he is promoting chaos and civil war," he fired back.

As a visiting professor at the American University in Cairo, which is a stronghold of Western liberalism, I often heard students express anger at America's invasion and occupation of Iraq, as well as its staunch support for Israel. "Bush has given Israel carte blanche to attack the Palestinians and Lebanese," Rania, a teenager with strikingly dark eyes, told me in the campus courtyard. "The war on terror is an open-ended war on Muslims," she insisted. Many students whom I interviewed at the American University in Beirut expressed similar views.

Around the same time as my visits to Al-Azhar, I attended an *iftar*, an evening meal after the daylong Ramadan fast, with hundreds of prominent Egyptians and Arabs of all political persuasions. The speaker, a moderate political leader and rising star in Egyptian society, said that that year's Ramadan coincided with a coordinated attack on Islam. I had not met a taxi driver, a fruit vendor, or a teacher who did not see a connection between the Danish cartoons portraying the Prophet Mohammed as a terrorist, President George W. Bush's use of the term "Islamo-fascism," and Pope

Benedict XVI's remarks linking Islam and violence. The speaker repeated the idea that the pope had given Bush religious justification for a war on Islam and Muslims, and guests nodded in agreement.

Of course, leading European countries such as France and Germany opposed the American venture in Iraq. The pope has also said that the U.S.-led invasion and occupation of Iraq is unjust, and opposed Israel's indiscriminate tactics against the Palestinians and the Lebanese. But this has not changed the minds of the many Muslims who see the West as united.

An Islamic leader, Abed al-Rahim Barakat, said to me, "President Bush himself used the word 'crusade' to describe his war on terror." "It was a slip of the tongue," I retorted. "No, it was a Freudian slip. He revealed what he feels deep inside," he said.

Five years after the September 11 attacks, Al Qaeda's notion of a clash of religions is no longer far-fetched. In both camps, tiny minorities are beating the drums, rallying the faithful to fight in a war they believe was caused by the other. Ordinary Muslims, not just Islamists and jihadists, view the "war on terror" as a war against their religion and values. Many Muslims who had initially condemned Al Qaeda and 9/11 are having second thoughts about bin Laden's fight against the Americans and their allies. Bin Laden has gained credibility in their eyes. "Now he is defending the ummah," a young rising poet, Massoud Hamed, confided to me.

American policy makers have little appreciation for how their military involvement in Iraq, as well as their staunch support of Israel, is radicalizing mainstream Muslim opinion and legitimizing radical groups that are waging armed struggles in Iraq, Palestine, Afghanistan, Lebanon, and elsewhere. *Al-Muqawama,* or "resistance," is the most popular slogan in the Muslim world today; it resonates deeply among men and women of all ages with religious and nationalist orientation alike. The plight of the Palestinians and Iraqis in particular echoes widely. I have yet to hear a Friday sermon in which believers are not reminded to lend a helping hand to their beleaguered Palestinian and Iraqi counterparts. The ability of Lebanon's Hizbollah, or "Party of God," to resist the Israeli military onslaught has lent credence to Al-Muqawama proponents. "Hizbollah's victory over the mighty Israeli army has broken the psychological barrier of fear among Muslims," a leading political activist told me. "We no longer fear American and Israeli military power. We are armed with faith." There is now widespread support for local jihad against foreign occupation of Muslim lands. If not for logistical barriers, tens of thousands of young Muslims would travel to Iraq, Palestine, and Lebanon to defend their brethren.

In fact, while Muslim public backing for global jihad is limited, Iraq today is one of the most promising theaters for the movement's revival. The American-led invasion and occupation of Iraq has given rise to a new generation of jihadists who differ dramatically from the first generation—the founding fathers—and the second generation of the Afghan Arabs or Al Qaeda.

Members of the first generation of the jihadist movement came from the middle class and upper-middle class and graduated from top scientific and social sciences departments in their countries' best universities. They possessed a complex, though distorted, grasp of various schools of Islamic thought and laid the theoretical foundation of jihadism utilized extensively by the Al Qaeda generation and the Iraq generation alike.

In contrast, many of the Iraq generation of jihadists come from the poverty belts of Arab and Muslim ghettos and streets—from the bottom of the heap, socially and economically. Many have shockingly little religious and formal education. I met teenagers who aspired to join the fight against the American occupiers and were nearly illiterate, with no grasp of interpretations of religious texts. They lacked the financial means—a few hundred U.S. dollars—to travel to Iraq, but those like them form a huge pool of potential recruits for global jihad that may be tapped into. This is an alarming development that has not been digested by American policy makers and their Iraqi allies.

Moreover, unlike the first and second generations, the Iraq jihadists do not make a clear distinction between the near enemy (Muslim ruling "renegades") and the far enemy (the United States and its allies). They are waging an all-out war against internal and external enemies alike. The lines of demarcation between Muslims and non-Muslims have also become blurred. The Iraq jihadists are willing to kill thousands of fellow Muslims who, in their eyes, are *kufar* ("apostates") and are at least as dangerous as Americans and Westerners.

In my conversation with members of the first generation and some of the Afghan Arabs, they were at a loss to explain the beastly acts of terror carried out by their Iraqi counterparts. While jihadists are conspiratorial by nature and ascribe all actions that are at odds with their conventional wisdom to Zionist and American plots, the Afghan Arabs conceded that the indiscriminate killings of Muslims and civilians was a by-product of the Iraq generation's scanty religious education and low social status.

"Foreign recruits to Iraq are vulnerable to Takfiri ideology [excommunication of Muslims and non-Muslims], which puts them on the wrong path," a leader of the first generation told me. "There is no other way to make sense of the proliferation of suicide bombings—thousands of operations—against fellow believers in Iraq."

In other words, foreign recruits to global jihad in Iraq are raw material easily molded by Al Qaeda leaders there. They serve as human bombs, carriers of death and destruction. We are witnessing further mutation and militarization with every jihadist generation. Who ever thought that the Iraq generation would be more violent than the Al Qaeda generation?

Both mainstream and militant Islamists whom I interviewed distanced themselves from the Iraq jihadists, who are "a liability to the honorable resistance," as one Islamist put it. However, the Iraq jihadists are no longer reliant on foreign recruits, but are quickly winning many Sunni Iraqis to their cause. In the last year a dramatic transformation has taken place within the Iraq generation with the homegrown Iraqi contingent outnumbering Arab recruits. The Iraq generation is becoming more Iraqi by the day.

Once the Americans exit Iraq, jihadists, who represent only about 5 percent of all fighters in Iraq, will face stiff opposition from the Sunni community that now provides them with protection and shelter. However, like parasites, Al Qaeda and its affiliates feed on turmoil and chaos. Al Qaeda is in the process of establishing an indigenous base in war-ravaged Iraq. If the country sinks into all-out war, it will become to global jihad what Afghanistan was in the 1990s and early 2000s. "America invaded Iraq under the pretext of [the U.S.] war against Al Qaeda," Kamal said, laughing. "But America succeeded in reviving a dormant Al Qaeda and the global jihad movement. It has also awakened the ummah from its political slumber."

The U.S. debate over Iraq focuses mainly on the effects of the American military presence on Al Qaeda and its affiliates—a tiny fringe in the political landscape in Iraq and beyond—while largely ignoring the negative effects of the war on mainstream Muslim opinion worldwide. The American military presence in the Arab heartland has become a liability, not an asset, to long-term political stability in Iraq and maintenance of the United States' vital interests in the region, particularly the fight against global jihad.

The sad irony is that the American war in Iraq has proved to be counterproductive to the struggle against the global jihad movement. It has given Al Qaeda central and its affiliates a new lease on life, and alienated the floating middle of Muslim opinion. Even U.S. intelligence services have finally and reluctantly arrived at the same conclusion.

NOTES

The main sources of this book are transcripts of hundreds of interviews and conversations I had with individuals throughout the Arab world between 1999 and 2005. The conversations were conducted in Arabic; the English translations are my own. Most of these individuals have been named, and the general dates and locations of the interviews are indicated in the text. Below are other sources that are either mentioned in each chapter or inform some of the material it contains.

I

PORTRAIT of a JIHADIST:
THE FIRST GENERATION

Memoirs and diaries I cite in this chapter were serialized in the Arab-language newspapers *Asharq al-Awsat, Al-Quds al-Arabi, Al Hayat,* and others.

Ayubi, Nazih N. *Political Islam: Religion and Politics in the Arab World.* London and New York: Routledge, 1991.

Gerges, Fawaz A. *The Far Enemy: Why Jihad Went Global.* New York: Cambridge University Press, 2005.

Kepel, Gilles. *Muslim Extremism in Egypt: The Prophet and Pharaoh.* Translated by Jon Rothschild. Berkeley and Los Angeles: University of California Press, 1985.

Mitchell, Richard P. *The Society of the Muslim Brothers.* Oxford: Oxford University Press, 1993.

Qutb, Sayyid. *Milestones.* Cedar Rapids, IA: The Mother Mosque Foundation, n.d.

———. *In the Shade of the Qur'an.* London: MWH, 1979.

II

THINGS FALL APART

El Khazen, Farid. *The Breakdown of the State in Lebanon: 1967–1976.* London: I.B. Tauris, 1999.

Fisk, Robert. *Pity the Nation: Lebanon at War.* Oxford: Oxford University Press, 1991.

Hamzeh, Nizar Ahmad. *In the Path of Hizbullah.* New York: Syracuse University Press, 2004.

Hudson, Michael C. *The Precarious Republic: Political Modernization in Lebanon.* Boulder: Westview Press, 1985.

Keddie, Nikki R., and Mark J. Gasiorowski, eds. *Neither East Nor West: Iran, the Soviet Union, and the United States.* New Haven, CT: Yale University Press, 1990.

Khalaf, Samir. *Lebanon's Predicament.* New York: Columbia University Press, 1987.

Salibi, Kamal. *The Modern History of Lebanon.* New York: Caravan Books, 1977.

————. *A House of Many Mansions: The History of Lebanon Reconsidered.* London: I.B. Tauris, 1988.

Wright, Robin. *Sacred Rage: The Wrath of Militant Islam.* New York: Simon and Schuster, 1985.

III

The WARRIORS of GOD: THE SECOND GENERATION

Memoirs and diaries I cite in this chapter were serialized in the Arab-language newspapers *Asharq al-Awsat, Al-Quds al-Arabi,* and *Al Hayat,* and others.

Messages to the World: The Statements of Osama bin Laden, edited and introduced by Bruce Lawrence. Translated by James Howarth. London and New York: Verso, 2005.

Benthall, Jonathan, and Jerome Bellion-Jourdan. *The Charitable Crescent: Politics of Aid in the Muslim World.* London: I.B. Tauris, 2003.

Bergen, Peter L. *Holy War, Inc.: Inside the Secret World of Osama Bin Laden.* New York: Free Press, 2001.

Fandy, Mamoun. *Saudi Arabia and the Politics of Dissent.* New York: St. Martin's Press, 1999.

Gerges, Fawaz A. *The Far Enemy: Why Jihad Went Global.* New York: Cambridge University Press, 2005.

National Commission on Terrorist Attacks Upon the United States. *The 9/11 Commission Report: Final Report of the National Commission on Terrorist Attacks Upon the United States.* New York: W. W. Norton, 2004.

Roy, Olivier. *Afghanistan: From Holy War to Civil War.* Princeton, NJ: Darwin Press, 1995.

Steven, Burg L., and Paul S. Shoup. *The War in Bosnia-Herzegovina: Ethnic Conflict and International Intervention.* Armonk, NY: M.E. Sharpe, 2000.

Seely, Robert. *Russo-Chechen Conflict 1800–2000: A Deadly Embrace.* Oxford: Frank Cass Publishers, 2001.

Smith, Sebastian. *Allah's Mountains: The Battle for Chechnya.* New York: I.B. Tauris 2001.

Stevenson, Jonathan. *Losing Mogadishu: Testing U.S. Policy in Somalia.* Annapolis: Naval Institute Press, 1995.

IV

The GREAT SATAN, NEAR AND FAR

Abdel-Malek, Kamal. *America in an Arab Mirror: Images of America in Arabic Travel Literature.* New York: Palgrave Macmillan, 2000.

Algar, Hamid. *Islam and Revolution: Writings and Declarations of Imam Khomeini.* Berkeley: Mizan Press, 1981.

Keddie, Nikki R., and Mark J. Gasiorowski, eds. *Neither East Nor West: Iran, the Soviet Union, and the United States.* New Haven, CT: Yale University Press, 1990.

Qutb, Sayyid. *Social Justice in Islam.* Translated by John B. Hardie. Oneonta, NY: Islamic Publications International, 2000.

———. *Islam: The Religion of the Future.* Riyadh, Saudi Arabia: International Islamic Federation of Student Organizations, 1984.

Rubin, Barry, and Judith Colp Rubin, eds. *Anti-American Terrorism and the Middle East: A Documentary Reader.* Oxford: Oxford University Press, 2002.

Wright, Robin. *Sacred Rage: The Wrath of Militant Islam*. New York: Simon and Schuster, 1985.

V

UNDER MIDDLE EASTERN EYES

Memoirs and diaries I cite in this chapter were serialized in the Arab-language publications *Asharq al-Awsat, Al-Quds al-Arabi, Al Hayat, Al-Mussawar* magazine, and others.

See the series of articles by Andrew Higgins and Alan Cullison in *The Wall Street Journal*—December 31, 2001; January 16, 2002; July 2 and 23, 2002; August 2 and 20, 2002; November 11, 2002; and December 20, 2002—discussing private messages and internal correspondence stored on Al Qaeda computers captured in Kabul immediately after the city fell to the pro-American Northern Alliance; a computer was acquired by Cullison, a reporter for the *Journal*. See also Cullison, Alan. "Inside Al-Qaeda's Hard Drive." *The Atlantic Monthly* 294, no. 2 (September 2004), 55–70.

Gerges, Fawaz A. *The Far Enemy: Why Jihad Went Global*. New York: Cambridge University Press, 2005.

Rubin, Barry, and Judith Colp Rubin, eds. *Anti-American Terrorism and the Middle East: A Documentary Reader*. Oxford: Oxford University Press, 2002.

VI

JIHAD DIASPORA

Memoirs and diaries I cite in this chapter were serialized in the Arab-language publications *Asharq Al-Awsat, Al-Quds al-Arabi, Al Hayat, Al-Mussawar* magazine, *Al Wasat,* and *Al Mustaqbal*. Western newspapers cited are the *New York Times, Washington Post, Los Ange-*

les *Times, Christian Science Monitor, Baltimore Sun, Guardian, Associated Press, Reuters, Agence France Press, Agence Global, Newsweek,* BBC, NPR, and ABC.com.

Brisard, Jean-Charles, and Damien Martinez. *Zarqawi: The New Face of Al-Qaeda.* New York: Other Press, 2005.

Dodge, Toby. *Inventing Iraq: The Failure of Nation-Building and a History Denied.* New York: Columbia University Press, 2003.

Gerges, Fawaz A. *The Far Enemy: Why Jihad Went Global.* New York: Cambridge University Press, 2005.

Hussein, Fuad. "Al-Zarqawi: The Second Generation of Al Qaeda," serialized in *Al-Quds al-Arabi,* May 13–30, 2004.

Napoleoni, Loretta. *Insurgent Iraq: Al-Zarqawi and the New Generation.* New York: Seven Stories Press, 2005.

EPILOGUE

al-Ashab, Mohammed. "Information Revealed by the Police in Rabat: European Cells to Recruit Moroccan Suicide Bombers and Send them to Iraq," *Al Hayat,* January 18, 2006.

———. "Testimonies of Moroccans Who Tried to Infiltrate Into Iraq: 'European' Networks to Smuggle 'Jihadists,'" *Al Hayat,* February 11, 2005.

———. "'European' Networks to Smuggle 'Jihadists' [Into Iraq]," *Al Hayat,* November 2, 2005.

Amin, Hazim. "Strangers Arrive During 'Al Qaeda Season' to Recruit Suicide Attackers to Iraq," part 1 out of 2, *Al Hayat,* January 26, 2006; part 2 out of 2, January 27, 2006.

Beeston, Richard. "Analysis: The Syrian Route to Jihad," *The Times* (London), June 21, 2005.

Beeston, Richard and James Hider, "Following the Trail of Death: How Foreigners Flock to Join the Holy War," *The Times* (London), June 25, 2005.

"Fundamentalists from 'Osbat al-Ansar' and the Salafi Current Leave Palestinian Camps in Lebanon to fight in Iraq," *Asharq al-Awsat*, October 4, 2005.

Hamidi, Ibrahim. "Syria Explains to Foreign Ambassadors What it Has Done to Prevent Infiltration into Iraq...," *Al Hayat*, July 21, 2005.

————. "'A Transnational Network' to Facilitate Transporting 'Jihadists' [Into Iraq]," *Al Hayat*, October 27, 2004.

MacFarquhar, Neil. "At Tense Syria-Iraq Border, American Forces Are Battling Insurgents Every Day," *The New York Times*, October 26, 2004.

McLean, Renwick. "20 Arrested as Spain Breaks Militant Networks," *The New York Times*, January 11, 2006.

Pitman, Todd. "North Africans Joining Iraq Islamic Fighters," *The Associated Press*, June 14, 2005.

Quinn, Patrick. "Most Suicide Bombs in Iraq by Foreigners," *The Associated Press*, June 30, 2005.

"Syria Says it Has Increased Security on Iraq Border, Detained Infiltrators," *The Associated Press*, October 28, 2005.

Whitlock, Craig. "In Europe, New Force for Recruiting Radicals," *The Washington Post*, February 18, 2005.

Zaatari, Mohammed. "Palestinian Refugees Joining Iraq Insurgency," *The Daily Star*, November 17, 2005.

GLOSSARY

adhan: call to prayer

al-Adou al-Baeed: far enemy (United States and its Western allies, particularly Israel)

al-Adou al-Qareeb: near enemy (apostate Muslim rulers)

al-Muqawama al-Iraqiya: Iraqi armed resistance

al-shabab: youth

al-solh: truce-making

asabiya: group or tribal solidarity

"Assalamu Aleykum": "peace be upon you"—a Muslim greeting

baiya: fealty

dar al-iman: abode of belief

dar al-kufr: abode of infidelity

da'wa: religious call

deen: religion

fard 'ayn: permanent and personal obligation

fard kifaya: collective duty

fatwa: religious edict or ruling

fiqh: Islamic jurisprudence

fitna: sedition

hakimiya: God's rule or sovereignty

hijra: emigration or exodus of the Prophet Mohammed from Mecca to Medina in 622

hizb al-din: Party of God

iman: belief or faith

Islamist: Muslim activist whose goal is to gain power and Islamize state and society; there are several kinds of Islamists—from moderate to mainstream to militant

jahili, jahiliya: designates the ignorance of Arabian society before the coming of Islam

jihad: struggle in the way of God, including armed struggle

jihadist: militant activist who rejects secular politics and is willing to use any means necessary to establish theocratic governments

keffiyeh: Palestinian headscarf worn by men

knafe: pastries composed of cheese, ground wheat, and rose water

kufar: infidels

kufr: infidelity

matawi'ah: erudite scholar

mujahedeen: Islamic fighters

mukhabarat: security apparatus

muktar: mayor

neksa: setback

qiyadi: endowed with leadership qualities

Qur'an: the book Muslims believe was revealed to Prophet Mohammed

sahaba: the companions of the Prophet

salafi: an ultra-conservative reformist who rejects intellectualism, mysticism, and sectarianism within Islam

shahid: martyr

shariah: Islamic law

Shiite: a Muslim who believes that the true successors to the Prophet should have been from Mohammed's family

shuja': courageous

shura: consultation

Sufi: an Islamic mystic

Sunnah: the recorded deeds and words of the Prophet, second in importance to the Qu'ran

Sunni: a Muslim who believes that the Prophet intended the community to select its own leader after his death

taghut: tyrant

tawba: repentance

ulema: religious scholars

ummah: Islamic community worldwide

"Wa Aleykum Assalam": "and peace be upon you too"—a Muslim greeting

Wahhabi or Wahhabism: a puritanical, religious sect founded by the eighteenth-century preacher Mohammed b. Abd al-Wahhab in Saudi Arabia

wahi: revelation

wilayet al-mar'ah: state ruled by a woman

SELECTED BIBLIOGRAPHY

Abdel-Malek, Kamal. *America in an Arab Mirror: Images of America in Arabic Travel Literature*. New York: Palgrave Macmillan, 2000.

Abu-Amr, Ziad. *Islamic Fundamentalism in the West Bank and Gaza: Muslim Brotherhood and Islamic Jihad*. Bloomington, IN: Indiana University Press, 1994.

Algar, Hamid, trans. *Islam and Revolution: Writings and Declarations of Imam Khomeini*. Berkeley: Mizan Press, 1981.

Arjomand, Said. *The Turban for the Crown: The Islamic Revolution in Iran*. New York: Oxford University Press, 1988.

Armstrong, Karen. *The Battle for God*. New York: Alfred A. Knopf, 2000.

Ayubi, Nazih N. *Political Islam: Religion and Politics in the Arab World*. London and New York: Routledge, 1991.

Baker, Raymond William. *Islam without Fear: Egypt and the New Islamists*. Cambridge, MA: Harvard University Press, 2003.

Bergen, Peter L. *Holy War, Inc.: Inside the Secret World of Osama Bin Laden*. New York: Free Press, 2001.

Bulliet, Richard. *The Case for Islamo-Christian Civilization*. New York: Columbia University Press, 2004.

Burgat, Francois. *Face to Face with Political Islam*. London: I.B. Tauris, 2003.

Burgat, Francois, and William Dowell. *The Islamic Movement in North Africa*. Austin, TX: University of Texas Press, 1997.

Clarke, Richard A. *Against All Enemies: Inside America's War on Terror*. New York: Free Press, 2004.

Coll, Steve. *Ghost Wars: The Secret History of the CIA, Afghanistan, and Bin Laden, from the Soviet Invasion to September 10, 2001*. New York: The Penguin Press, 2004.

Eickelman, Dale F., and James Piscatori. *Muslim Politics*. Princeton, NJ: Princeton University Press, 1996.

El-Affendi, Abdelwahab. *Turabi's Revolution: Islam and Power in Sudan*. London: Grey Seal, 1991.

Enayat, Hamid. *Modern Islamic Political Thought*. London: Palgrave Macmillan, 1982.

Entelis, John. *Algeria: The Revolution Institutionalized*. Boulder, CO: Westview Press, 1986.

Esposito, John L., ed. *Voices of Resurgent Islam*. New York: Oxford University Press, 1983.

Esposito, John L., and John O. Voll. *Makers of Contemporary Islam*. Oxford: Oxford University Press, 2001.

Fanon, Frantz. *The Wretched of the Earth*. Translated by Constance Farrington. New York: Grove Press, 1963.

Gerges, Fawaz A. *The Far Enemy: Why Jihad Went Global.* New York: Cambridge University Press, 2005.

———. *America and Political Islam: Clash of Cultures or Clash of Interests?* New York: Cambridge University Press, 1999.

Hourani, Albert. *A History of the Arab Peoples.* Cambridge, MA: Belknap Press of Harvard University Press, 1991.

———. *Arabic Thought in the Liberal Age: 1798–1939.* New York: Cambridge University Press, 1983.

Hroub, Khaled. *Hamas: Political Thought and Practice.* Beirut, Lebanon: Institute for Palestinian Studies, 2000.

Kepel, Gilles. *Muslim Extremism in Egypt: The Prophet and Pharaoh.* Translated by Jon Rothschild. Berkeley and Los Angeles: University of California Press, 1985.

———. *The War for Muslim Minds: Islam and the West.* Translated by Pascale Ghazaleh. Cambridge, MA: Belknap Press of Harvard University Press, 2004.

Khaldun, Ibn. *The Muqaddimah: An Introduction to History.* Translated by Franz Rosenthal. Princeton, NJ: Princeton University Press, 1967.

Lacoste, Yves. Translated by David Macey. *Ibn Khaldun: The Birth of History and the Past of the Third World.* London: Verso, 1984.

Maalouf, Amin. *The Crusades Through Arab Eyes.* Translated by Jon Rothschild. New York: Schocken Books, 1989.

Malcolm, Noel. *Bosnia: A Short History.* New York: New York University Press, 1994.

Marty, Martin E., and R. Scott Appleby. *Fundamentalisms Observed.* Chicago: Chicago University Press, 1991.

Mitchell, Richard P. *The Society of the Muslim Brothers*. Oxford: Oxford University Press, 1993.

Mottahedeh, Roy. *The Mantle of the Prophet: Religion and Politics in Iran*. New York: Simon and Schuster, 1985.

Nasr, Seyyed Vali Reza. *The Vanguard of the Islamic Revolution: The Jama'at-i Islami of Pakistan*. Berkeley: University of California Press, 1994.

———. *Mawdudi and the Making of Islamic Revolution*. New York: Oxford University Press, 1996.

National Commission on Terrorist Attacks Upon the United States. *The 9/11 Commission Report: Final Report of the National Commission on Terrorist Attacks Upon the United States*. New York: W. W. Norton, 2004.

Pape, Robert A. *Dying to Win: The Strategic Logic of Suicide Terrorism*. New York: Random House, 2005.

Peters, Rudolph. *Jihad in Classical and Modern Islam: A Reader*. Princeton, NJ: Markus Wiener Publishers, 1996.

Pierre, Andrew J., and William B. Quandt. *The Algerian Crisis: Policy Options for the West*. Washington, DC: Carnegie Endowment for International Peace, 1996.

Qutb, Sayyid. *Milestones*. Cedar Rapids, IA: The Mother Mosque Foundation, n.d.

———. *In the Shade of the Qur'an*. London: MWH, 1979.

———. *Social Justice in Islam*. Translated by John B. Hardie. Oneonta, NY: Islamic Publications International, 2000.

———. *Islam: The Religion of the Future*. Riyadh, Saudi Arabia: International Islamic Federation of Student Organizations, 1984.

Rashid, Ahmed. *Taliban: Militant Islam, Oil and Fundamentalism in Central Asia.* New Haven, CT: Yale University Press, 2000.

Rodinson, Maxime. *Islam and Capitalism.* Translated by Brian Pearce. New York: Random House, 1974.

Roy, Olivier. *Globalized Islam: The Search for a New Ummah.* New York: Columbia University Press, 2004.

————. *The Failure of Political Islam.* Translated by Carol Volk. Cambridge, MA: Harvard University Press, 1994.

Rubin, Barry, and Judith Colp Rubin, eds. *Anti-American Terrorism and the Middle East: A Documentary Reader.* Oxford: Oxford University Press, 2002.

Sells, Michael Anthony. *The Bridge Betrayed: Religion and Genocide in Bosnia.* Berkeley: University of California Press, 1998.

Shahin, Emad Eldin. *Political Ascent: Contemporary Islamic Movements in North Africa.* Boulder, CO: Westview Press, 1998.

Tripp, Charles. *A History of Iraq.* Cambridge, England: Cambridge University Press, 2002.

Weber, Max. *The Sociology of Religion.* Translated by Ephraim Fischoff. Boston: Beacon Press, 1963.

Wright, Robin. *Sacred Rage: The Wrath of Militant Islam.* New York: Simon and Schuster, 1985.

Zubaida, Sami. *Islam, the People and the State: Essays on Political Ideas and Movements in the Middle East.* London and New York: Routledge, 1989.

ACKNOWLEDGMENTS

I COULD NOT HAVE written *Journey of the Jihadist* without the cooperation of hundreds of Islamic activists and militants. Trusting that I would tell their stories truthfully and honestly, many gave me hours and hours of their time. Listing them all, in order to recognize their vital contribution to this work, would be impossible. But I need to acknowledge Kamal el-Said Habib, who in some ways is the main character of this book and certainly its central source. Soon after meeting him in 1999 I realized that his journey was emblematic of that of so many other jihadists. Kamal was generous with his time, helping me retrace his steps from his student days to the present, patiently answering my endless enquiries. He furnished me with some of his unpublished manifestos, dating from his early days as a revolutionary. This was courageous of him. He spent ten years in prison, was rearrested several times, and lives still with the threat that he may be arrested yet again. Kamal also introduced me to many of his former comrades, who shared their own experiences and perspectives. They will all likely disagree with many of my conclusions. But I hope they also feel that I have not distorted their viewpoints and that, to some degree at least, I have let them tell their own stories.

My editor, Tim Bent, pushed me hard to go beyond the confines of academic research, and to seek and reveal the human face of a struggle that has taken so many lives. I hope the results do justice to that effort. Ayesha Pande helped to structure the book from its early stages onward. Her critical feedback greatly improved it. Frank Browning also improved much of the material. His input was far more substantive than style, and I'm indebted to him. My colleague, Dr. Kristin Sands, read the revised manuscript and offered important suggestions. Jennifer Lyons went beyond the call of duty to bring this book to life.

A MacArthur Foundation fellowship and a Smith Richardson Foundation grant enabled me to spend two years in the Middle East. I am grateful for the opportunities they afforded.

I dedicate this book to my family, whose love and support have nourished and sustained me. I could not have written it without my wife, Nora's, generosity of spirit and intellectual rigor. My children, Bassam, Annie-Marie, Hannah, and Laith were with me every minute of the journey. They inspire me to search deeper and harder for the ties that bind us all.

INDEX

Jama'at al-Tawhid wa'al-Jihad, 252

Jamaat-i-Islami, 119, 177

jihad. See also Al Qaeda; jihadists
and jihadism; mujahedeen
in Afghanistan, 109–13, 119–24
bin Laden and, 1–2, 87, 107–8
as eternal revolution, 35–36
goals in Iraq, 255
Muslim neglect of, reprimands
for, 6–7
Qur'an and, 2–3, 87, 106, 179,
238, 272
as response to U.S. war against
Al Qaeda, 1–2

al-Jihad, 19
former activists of, 30
renamed Tanzim al-Jihad, 21

Jihad Group, 224

jihadists and jihadism. *See also*
Islamists; mujahedeen; Muslim
Brotherhood
in Afghanistan, 109–13, 119–24,
172–76
Al Qaeda versus, 21–23, 90–91
American foreign policy and,
166–67, 223
Arab-Israeli war (1967) and, 11,
32–33
at Cairo University, 21, 30–34, 41
changes within, 59, 235–40,
251–52, 255–56, 261–72
core in Saudi Arabia, 105–6
criticisms of September 11, 2001
terrorist attacks, 211–29
culture of death and, 5
culture of political realism and,
16–17

democratic transformation and,
15–16, 53–54
in Egypt, 21–23, 35–60
elite leadership of, 39
first generation, 19–60, 97, 112,
114, 115
as global offensive, 102–25,
173–75, 180–81, 220–25, 266–67
goal of modern, 3–4, 175–76
human dimensions of, 30
impact of September 11, 2001
terrorist attacks on, 202–4,
212–13
interviews with jihadists, 13–18.
See also Abu-Jandal (Nasir
Ahmad Nasir Abdallah al-
Bahri); Habib, Kamal el-Said
at Ain al-Hilweh (Palestinian
refugee camp), 191, 196–97,
198–202
in Sidon, 191, 196–98
in Tripoli, 191, 192–96
Islamists versus, 11–18
itinerary of, 273–77
journey of, 1–3
key to understanding, 11–12
lack of vision, 55–56
in Lebanon, 61–62, 67–92,
191–202
martyrdom and, 35–41, 47–51,
60, 87–88
motivation for, 13–14
near enemy versus far enemy
and, 4, 11, 165–66, 169, 170,
176–77, 180–81, 229
neo-fundamentalism and, 16–17
new consensus among, 239–41

Omar, Mullah, 137, 204–6, 210
Omar, Omar Mahmoud Abu, 224
Osbat al-Ansar (League of the
    Partisans), 197
Osbat al-Nour (League of Light),
    197
Ottoman Empire
    art and literature of, 42
    fall of, 7, 42–43, 97, 244

Pakistan, 109, 118–21, 137, 173–75,
    177, 254
    Services Bureau, 109, 119, 120–21,
    123, 174
    support for Afghanistan,
    109–10
Palestine
    bin Laden and, 254
    conflict with Israel, 7, 24, 49, 62,
    85–86, 101–2, 159–63, 176, 187,
    190–91, 200, 226, 235, 242, 250,
    254, 255
    Hamas victory in 2006, 9–10
    jihad in, 122
    partitioning of, 159–61
    refugees in Lebanon, 62–63, 67,
    68–76, 191, 196–97, 198–202
    U. S. policy toward, 184
Palestine Liberation Organization
    (PLO), 67, 68–69, 78, 199
Pentagon attack (2001). See
    September 11, 2001 terrorist
    attacks
Phalangists, 62, 64, 71, 75–76
Philippines, 107–8, 116, 121–22
Protocols of the Elders of Zion, 7
Putin, Vladimir, 124–25

Qaddafi, Mu'ammar, 85–86
Qardawi, Yusuf, 204, 240, 242
Qatada, Abu, 224
Qazai, Saraj, 266
qiyadi, 38
Qoleilat, Ibrahim, 67, 78
Al-Quds al-Arabi (newspaper), 95,
    129–30, 213, 259
Qur'an, 159. See also Islam
    jihad and, 2–3, 87, 106, 179, 238,
    272
    terrorism and, 11–12
Qutb, Sayyid, 25, 35–40, 46, 47,
    122, 203
    execution of, 35, 37, 38
    U.S. and, 36–37, 144–46, 149–55,
    156–61, 162–63, 170–71

Rabbani, Mohammed, 140–41
Ramadan, 28, 52, 236
Ramadan, Mustapha Darwish,
    273–77
Ramírez Sánchez, Ilich (the Jackal),
    88
Reagan, Ronald, 89, 99, 114, 164,
    171, 172
religion. See also Christianity;
    Islam; Judaism
    as end in itself, 33–38
    sociopolitical goals behind,
    10–11
    as tool of dissent, 9
Revolutionary Guard, 85
Rida, Rashid, 23
Al-Risala (magazine), 36–37, 160
Royal Institute of International
    Affairs, 267